North Carolina Test Prep Workbook for Holt Middle School Math, Course 1 Teacher's Edition

Help for Grade 6 EOG

HOLT, RINEHART AND WINSTON

A Harcourt Education Company

Orlando • **Austin** • New York • San Diego • Toronto • London

Printed in the United States of America

ISBN 0-03-035854-X

3 179 05 04

CONTENTS

Question	Item Objective	Lesson in Student Book	North Carolina Standards
1.	Represent numbers by using exponents.	1–3	1.05
2.	Choose an appropriate method of computation.	1–6	1.07
3.	Choose an appropriate method of computation.	1–6	1.07
4.	Estimate with whole numbers.	1–2	1.04c
5.	Use order of operations.	1–4	1.04
6.	Find, recognize, describe, and extend patterns in sequences.	1–7	1.03
7.	Estimate with whole numbers.	1–2	1.04c
8.	Find, recognize, describe, and extend patterns in sequences.	1–7	1.03
9.	Use order of operations.	1–4	1.04
10.	Find, recognize, describe, and extend patterns in sequences.	1–7	1.03
11.	Compare and order whole numbers using place value or a number line.	1–1	1.03
12.	Represent numbers by using exponents.	1–3	1.05
13.	Estimate with whole numbers.	1–2	1.04c
14.	Translate between words and math.	2–2	5.01
15.	Identify and evaluate expressions.	2–1	5.02
16.	Identify and evaluate expressions.	2–1	5.02
17.	Determine whether a number is a solution of an equation.	2–3	5.02, 5.03
18.	Solve whole-number multiplication equations.	2–6	1.04
19.	Identify and evaluate expressions.	2–1	5.01, 5.02
20.	Solve whole number addition equations.	2–4	1.04
21.	Solve whole number division equations.	2–7	1.04
22.	Determine whether a number is a solution of an equation.	2–3	5.01, 5.02, 1.04
23.	Solve whole-number multiplication equations.	2–6	1.04
24.	Determine whether a number is a solution of an equation.	2–3	5.01, 5.02, 1.04
25.	Identify and evaluate expressions.	2–1	5.01, 5.02
26.	Solve whole-number multiplication equations.	2–6	5.02, 5.03
27.	Identify and evaluate expressions.	2–1	5.01, 5.02, 5.03
28.	Multiply decimals by whole numbers and by decimals.	3–6	1.04
29.	Write, compare, and order decimals using place value and number lines.	3–1	1.03
30.	Write large numbers in scientific notation.	3–5	1.06
31.	Multiply and divide decimals by powers of ten and convert metric measurements.	3–4	1.05, 1.07
32.	Divide decimals by whole numbers.	3–7	1.04
33.	Solve problems by interpreting the quotient.	3–9	1.04
34.	Write, compare, and order decimals using place value and number lines.	3–1	1.03
35.	Multiply and divide decimals by powers of ten and convert metric measurements	3–4	1.04
36.	Solve equations involving decimals.	3–10	5.03
37.	Write, compare, and order decimals using place value and number lines.	3–1	1.03
38.	Add and subtract decimals.	3–3	1.04

Assessment in Standardized Test Prep Workbook

Diagnostic Test

The Diagnostic Test is a multiple choice test that covers objectives found on most standardized tests.

This test, on pages 1-24 of the Test Prep Workbook, can be used to measure student achievement before instruction. Each test item is correlated to a lesson in the student textbook where this content is taught. Results of this test can be used to create individual remediation plans.

The sample of test items is large enough that careless errors do not mask students' knowledge.

The performance observed on a specific test item is indicative of how the student will perform on a test with similar test items.

Test Taking Strategies

There are strategies for taking standardized tests given for every chapter. These strategies include working backwards, finding patterns, knowing how tests are scored, eliminating answers to multiple choice problems, knowing how to grid in answers, and drawing diagrams.

Chapter Standardized Tests

There is a two-page test for every chapter. Each test is comprised of multiple choice, gridded response, short response, and extended response questions. The questions cover the content of the chapter. There are answer sheets for each test.

Practice Tests

There are two practice standardized tests. Each test mimics the format of the state test.

Other Materials Available for Standardized Test Prep

Countdown Transparencies

There are 32 transparencies that can be used as lesson warm ups to help your students prepare for standardized tests. Each transparency has two multiple choice problems that are similar to the types of problems found on standardized tests.

Standardized Test Prep Video

A video is available that can be used to help your students prepare for standardized tests. The video has five questions for each chapter. Two of the questions have worked out solutions to help students understand one way of solving a problem.

State Test Prep CD-ROM

The *State Test Prep* CD-ROM provides practice items to help students prepare for standardized tests. This tool enables you to quickly create printed tests, Internet tests, and LAN-based tests. You can use sample chapter tests or customize your own tests from the multiple test banks provided.

Chapter tests include items similar to those found on your state test. Dynamic questions (algorithms) on the sample chapter tests allow you to generate and print more than one version of each test.

The content for each test bank is arranged by state standards. Tests from the test banks can be generated by one of five methods. You can select questions randomly, from a list, while you are viewing questions, by criteria such as specific state standards, or you can select all of the questions. As in the sample chapter tests, the test banks include dynamic questions.

Administering the Test

Students will need a copy of the test, separate answer sheets, and scratch paper. Blackline masters of the test and answer sheets are provided in this Manual.

The Diagnostic Test is not a timed test. Students should be given ample time to respond to all test items. You may want to give the test over several days.

This test was designed to be used without a calculator.

Directions for Administering the Test

1. Explain the purpose of the test and why it is being given. Briefly describe the test format and the time allotted to take the test.
2. Inform students that calculators are not permitted for the test.
3. Have students examine the answer sheet as you discuss how they are to record their answers.
4. Ask if there are any questions and respond to them.
5. Circulate around the room. If a student appears to be stuck on a test item, encourage the student to move on with the other test items and return to this item later.

Class Profile for Diagnostic Test

The Class Profile can be used to summarize the results of the Diagnostic Test. You can use the Class Profile to make better use of your classroom instruction time. For example, if the test results indicate that most students are proficient in decimals, you may decide to present a brief unit on decimals in class and provide remedial work for weak students. However, if the results indicate that most students are not proficient in fractions, you may choose to spend more classroom instruction time on this topic.

Interpreting the Diagnostic Test Results

The scores for an individual student can be summarized for each of the 12 chapters.

CHAPTER	TEST ITEMS	PROFICIENCY LEVEL	STUDENT SCORE	PROFICIENCY SCORE (%)
Number Toolbox	1-13	10/13	/13	
Introduction to Algebra	14-27	10/14	/14	
Decimals	28-40	10/13	/13	
Number Theory and Fractions	41-53	10/13	/13	
Fraction Operations	54-65	9/12	/12	
Collect and Display Data	66-76	8/11	/11	
Plane Geometry	77-87	8/11	/11	
Ratio, Proportion, and Percent	88-101	10/14	/14	
Integers	102-115	10/14	/14	
Perimeter, Area, and Volume	116-126	8/11	/11	
Probability	127-138	9/12	/12	
Functions and Coordinate Geometry	139-149	8/11	/11	

By referring to the Proficiency Score column, you can quickly determine whether the student has failed to demonstrate proficiency in one or more chapters.

Using the Class Profile

The Class Profile can be used to summarize the results of the Inventory Test. By using the Class Profile, you can make better use of classroom instruction time. For example, if the test results indicate that most students are proficient in decimals, you may decide to present a brief unit on decimals and provide remedial work for weak students. However, if the results also indicate that most students are not proficient in fractions, you may choose to spend more time teaching that chapter.

Class Profile for Diagnostic Test

STUDENT NAME	Ch 1	Ch 2	Ch 3	Ch 4	Ch 5	Ch 6	Ch 7	Ch 8	Ch 9	Ch 10	Ch 11	Ch 12
1.												
2.												
3.												
4.												
5.												
6.												
7.												
8.												
9.												
10.												
11.												
12.												
13.												
14.												
15.												
16.												
17.												
18.												
19.												
20.												
Total Number Proficient												
% of Students Proficient												

Holt Middle School Math　xiii

Class Profile for Diagnostic Test

STUDENT NAME	Ch 1	Ch 2	Ch 3	Ch 4	Ch 5	Ch 6	Ch 7	Ch 8	Ch 9	Ch 10	Ch 11	Ch 12
21.												
22.												
23.												
24.												
25.												
26.												
27.												
28.												
29.												
30.												
Total Number Proficient												
% of Students Proficient												

The Class Profile will help you look for patterns of performance among students.

Some ways you may choose to use the Class Profile include:

- grouping students with similar instructional needs
- identifying content for extra teaching
- identifying skills that need regular reinforcement
- noting whether students' class performance verifies strengths and weaknesses identified on the test

Standardized Test Practice

Class Record Form

CHAPTER TESTS

School												
Teacher												
NAMES	Date											

Answer Grids

Holt Middle School Math

Diagnostic Test

1. What is the value of 4^2?

A 2   **C** 8

B 4   **D** 16

2. Sandi can buy four CDs for $24.00. Which operation would she use to find the price of one CD?

F addition

G subtraction

H multiplication

I division

3. The table shows the attendance at a professional basketball team's first four games of the season.

Game 1	Game 2	Game 3	Game 4
46,324	47,112	49,518	48,301

Which method would you use to find the average attendance at the basketball team's first four games of the season?

A paper and pencil

B mental math

C models

D calculator

4. Each student in the sixth grade brought in $3.75 for the field trip to the museum. If there are 231 students in sixth grade, about how much money was brought in for the trip?

F $600

G $700

H $800

I $1,100

5. Use the order of operations to find the value of the expression.

$58 + 8(45 \div 9) - 21$

A 42   **C** 225

B 77   **D** 309

6. Which two numbers come next in the pattern?

0, 3, 7, 12, 18, …

F 25, 33

G 25, 32

H 28, 33

I 27, 34

7. A section of the theater has been reserved for 378 people. If each row in the theater seats 18 people, about how many rows have been reserved?

A 15 rows

B 18 rows

C 20 rows

D 25 rows

8. Keith has a bank account. Each week he deposits more than he did the week before. His deposits are listed below. Following this pattern, how much will he deposit next week?

Week 1	Week 2	Week 3	Week 4	Week 5
$15.00	$28.00	$43.00	$60.00	?

F $68.00

G $72.00

H $79.00

I $83.00

Holt Middle School Math   Course 1

Diagnostic Test

9. Use the order of operations to find which expression is true.

A $8 + 12(14 - 4) + 6 = 206$

B $10 - 8 \times 6 + 7 = 26$

C $14 + 6(7 - 3) - 8 = 30$

D $4 + 5(6 + 2) - 5 = 66$

10. Identify the best rule for the pattern.

4	8	16	32	64

F The numbers are increasing by 4.

G The numbers are increasing by 8.

H The numbers are doubling.

I The numbers are tripling.

11. Which set of numbers are ordered from least to greatest?

A 14,342; 13,449; 14,288

B 5,678; 5,687; 5,768

C 56,940; 57,981; 57,891

D 1,343; 1,342; 1,234

12. What whole number when raised to the third power equals 64?

F 2 **H** 6

G 4 **I** 8

13. A local bank has thirty thousand dollars at the beginning of the day. During the day $19,761 is withdrawn and $31,012 is deposited. Estimate how much money the bank has at the end of the day.

A $20,000

B $30,000

C $40,000

D $50,000

14. Which equation represents the following statement?

A number is multiplied by the sum of six and eight. The answer is four hundred thirty-two.

F $n(6 \times 8) = 432$

G $8 + 6 \times n = 432$

H $n \times (6 + 8) = 432$

I $8n + 6 = 432$

15. What is the value of the expression $15s - 4t$ when $s = 6$ and $t = 2$?

A 16

B 19

C 72

D 82

16. Quick Clean is a house cleaning service. The service charges $15 per room plus $12 per hour after the first hour. Which expression describes how to find the total cost?

F $15r + 12h$

G $15r + 12(h - 1)$

H $15r + 12(h + 1)$

I $15r(12h)$

17. Which of the following equations has a solution of 6?

A $x + 13 = 26$

B $x + 24 = 32$

C $x + 42 = 50$

D $x + 15 = 21$

Holt Middle School Math **Course 1**

Diagnostic Test

18. A mechanic shop uses the following formula to determine the total cost for auto repairs.

Cost = \$85 + 55h, where h is the number of hours that the repair takes.

If the cost of a repair was \$525, how many hours did the repair take?

F 6 hours

G 7 hours

H 8 hours

I 9 hours

19. Louie needs 25 more hours of training to become a lifeguard. He has already received 8 hours of training. Which equation shows the total number of training hours Louie needs to become a lifeguard?

A $25 + x = 8$

B $x - 8 = 25$

C $8 - x = 25$

D $25 - x = 8$

20. Sam has six blue shirts, w white shirts, and four red shirts. How many white shirts does he have if he has 15 shirts in all?

F 5 shirts

G 6 shirts

H 10 shirts

I 11 shirts

21. What is the value of d in the following equation?

$$\frac{d}{4} = 12$$

A $d = 3$ **C** $d = 24$

B $d = 16$ **D** $d = 48$

22. Which of the following equations does NOT have a solution of $x = 4$?

F $13x = 52$

G $x - 6 = 2$

H $x + 16 = 20$

I $\frac{x}{2} = 2$

23. Jessie wants to solve the equation $4x = 112$. What step should she take first?

A Add 4 to both sides.

B Subtract 4 from both sides.

C Multiply 4 to both sides.

D Divide by 4 on both sides.

24. Which of the following is a solution to the equation $x + 6 = 24$?

F $x = 4$

G $x = 18$

H $x = 30$

I $x = 36$

Holt Middle School Math   Course 1

Diagnostic Test

25. Which of the following describes the expression $5x + 8$?

    A five more than eight times a number

    B eight more than the product of five and a number

    C the sum of a number and eight

    D the product of a number and five

26. Which of the following is the solution to the equation $15x = 225$?

    F $x = 10$

    G $x = 12$

    H $x = 15$

    I $x = 18$

27. Which expression is the rule for the data in the table?

h	Data
3	16
5	26
7	36

    A $h + 13$

    B $5h + 1$

    C $4h + 6$

    D $6h - 3$

28. Kevin makes $13.85 an hour. Last week he worked 40 hours. How much did Kevin earn last week?

    F $54.40

    G $554.00

    H $5,540.00

    I $55,400.00

29. Bob recorded the average ocean temperatures for the past four weeks. Order these temperatures from least to greatest.

Week 1	Week 2	Week 3	Week 4
83.08°	82.19°	82.09°	82.21°

    A 82.09°, 82.19°, 82.21°, 83.08°

    B 82.09, 83.08°, 82.19°, 82.21°

    C 82.21°, 82.19°, 82.09°, 83.08°

    D 83.08°, 82.21°, 82.19°, 82.09°

30. The Mississippi River runs 2,340 miles. Which of the following correctly shows this number written in scientific notation?

    F 23.4×10^2

    G 2.34×10^2

    H 2.34×10^3

    I 0.234×10^4

31. Brandon went to the doctor for his check-up. He weighed 46.82 kg and was 1.39 m tall. What is his weight in grams?

    A 468.2 g

    B 1,390 g

    C 4,682 g

    D 46,820 g

32. Sam bought six movie tickets for $34.50. How much did one ticket cost?

    F $5.25

    G $5.50

    H $5.75

    I $6.00

Holt Middle School Math   Course 1

Diagnostic Test

33. Karen needs to purchase paper cups for the student council breakfast. There are 12 cups in a package for $1.65. If there are 115 students and parents attending the breakfast, how many packages of cups does Karen need to purchase?

A 7

B 8

C 9

D 10

34. Which decimal represents the shaded portion of the figure?

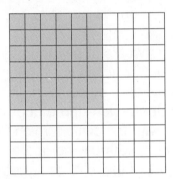

F 0.06

G 0.36

H 0.60

I 0.64

35. Molly walked 4.18 km, Lindsay walked 4,181 m, Shannon walked 4,111 m, and Beth walked 4.109 km. Who walked the farthest?

A Molly

B Lindsay

C Shannon

D Beth

36. Which value of y makes the equation true?

$45.092 + y = 78.1$

F $y = 33.008$

G $y = 33.8$

H $y = 122.12$

I $y = 123.19$

37. Which of these decimals is smaller than 0.036?

A 0.009   **C** 0.101

B 0.052   **D** 0.200

38. Add.

$23.009 + 4.09 + 0.287 = ?$

F 27.099

G 27.386

H 30.779

I 314.099

39. Mary's pace measures 35 cm. If she walks 2,000 paces, how many meters did she walk?

A 0.7 m

B 7.0 m

C 70.0 m

D 700.0 m

40. Which is the best estimate for $678.70 + 148.09$?

F 700

G 800

H 850

I 900

Diagnostic Test

41. Terri has 36 bracelets, 18 necklaces, and 27 rings. If she puts the same number of each type of jewelry in each box, how many boxes will she have?

A 4   **C** 6

B 5   **D** 9

42. Amy is thinking of the smallest number that is divisible by 10, 12, and 15. What is the number?

F 30

G 60

H 90

I 120

43. What is the prime factorization of 245?

A 7^3

B $7^2 \times 5$

C $3 \times 2^3 \times 5$

D 7×35

44. Which of the following fractions is NOT equivalent to $\frac{3}{8}$?

F $\frac{12}{32}$

G $\frac{9}{27}$

H $\frac{15}{40}$

I $\frac{6}{16}$

45. Which of the following is a prime number?

A 39

B 109

C 121

D 147

46. Sandi lost $1\frac{3}{8}$ pounds during basketball conditioning. Express this fraction as a decimal.

F 0.375 pounds

G 1.3 pounds

H 1.375 pounds

I 1.40 pounds

47. Brenda works at a veterinary clinic. One of her duties is to weigh the animals as they come in. In the first hour, there were three dogs that came into the clinic. Which of the following correctly compares the weights of these dogs?

A 5.65 kg $< 5\frac{2}{5}$ kg $< 5\frac{3}{4}$ kg

B 5.65 kg $< 5\frac{3}{4}$ kg $< 5\frac{2}{5}$ kg

C $5\frac{3}{4}$ kg $< 5\frac{2}{5}$ kg < 5.65 kg

D $5\frac{2}{5}$ kg < 5.65 kg $< 5\frac{3}{4}$ kg

Holt Middle School Math   **Course 1**

Diagnostic Test

48. Carol had $5\frac{7}{8}$ cups of flour. She used $3\frac{5}{8}$ cups of flour for three fruit cakes. How much flour does she have left?

F $2\frac{1}{8}$ cups

G $2\frac{3}{16}$ cups

H $2\frac{1}{4}$ cups

I $2\frac{3}{4}$ cups

49. Glenn spends 105 minutes exercising every day. He spends $\frac{3}{5}$ of the time running. How much time does Glenn spend running?

A 57 minutes

B 60 minutes

C 63 minutes

D 66 minutes

50. Jamie needs $6\frac{1}{6}$ yards of silk and $2\frac{5}{6}$ yards of lace to make a dress. How much total material does she need to buy?

F $8\frac{1}{2}$ yards

G 9 yards

H $9\frac{1}{2}$ yards

I 10 yards

51. To make lemonade, Marcie adds $2\frac{1}{4}$ cups of sugar to a gallon of water with six lemons. How much sugar will Marcie need if she triples her recipe?

A 6 cups

B $6\frac{1}{4}$ cups

C $6\frac{1}{2}$ cups

D $6\frac{3}{4}$ cups

52. What is the decimal equivalent to the fraction $\frac{1}{2}$?

F 0.05

G 0.2

H 0.5

I 1.2

53. Which pair of fractions are equivalent?

A $\frac{5}{9}$ and $\frac{20}{35}$

B $\frac{16}{28}$ and $\frac{4}{7}$

C $\frac{13}{39}$ and $\frac{1}{4}$

D $\frac{25}{54}$ and $\frac{5}{15}$

54. What is $\frac{3}{4}$ of $\frac{16}{21}$?

F $\frac{8}{15}$

G $\frac{4}{7}$

H $\frac{63}{64}$

I $\frac{8}{7}$

Holt Middle School Math   Course 1

Diagnostic Test

55. It took Stephen $3\frac{1}{2}$ hours to clean the pool. He spent $\frac{2}{3}$ of the time filling the pool with clean water. How long did it take Stephen to fill the pool?

A $1\frac{2}{3}$ hours

B 2 hours

C $2\frac{1}{3}$ hours

D 3 hours

56. Maggie runs $4\frac{1}{4}$ miles every evening. She has a goal of running a total of 221 miles. How many evenings does Maggie need to run to reach her goal?

F 21 evenings

G 52 evenings

H 510 evenings

I 940 evenings

57. Phil, Maddie, and Tom ordered one large family salad. Phil ate $\frac{1}{4}$ of the salad. Maddie ate $\frac{5}{8}$ of the salad. How much salad was left for Tom to eat?

A $\frac{1}{2}$ of the salad

B $\frac{1}{4}$ of the salad

C $\frac{1}{8}$ of the salad

D $\frac{1}{12}$ of the salad

58. In December, a severe snowstorm struck Tara's hometown. She measured and recorded the amount of snow that collected on the sidewalk.

Mon.	Tues.	Wed.	Thurs.	Fri.
$6\frac{1}{2}$ in.	$8\frac{1}{4}$ in.	$2\frac{3}{8}$ in.	$1\frac{3}{4}$ in.	$5\frac{1}{2}$ in.

What was the total amount of snow that fell in five days?

F $19\frac{1}{2}$ in.

G $21\frac{3}{4}$ in.

H $24\frac{3}{8}$ in.

I $26\frac{7}{8}$ in.

59. What is the least common multiple of 3, 5, and 8?

A 15

B 40

C 120

D 240

60. When Nick entered the sixth grade, he was $58\frac{1}{2}$ inches tall, which is $4\frac{3}{4}$ inches taller than he was when he entered fifth grade. How tall was Nick when he entered fifth grade?

F $52\frac{3}{4}$ inches

G $53\frac{1}{4}$ inches

H $53\frac{3}{4}$ inches

I $54\frac{1}{4}$ inches

Holt Middle School Math   Course 1

Diagnostic Test

61. What is $1\frac{4}{5} \div \frac{9}{20}$?

  A $\frac{1}{4}$

  B $\frac{4}{9}$

  C $\frac{9}{20}$

  D 4

62. What is the reciprocal of $4\frac{6}{9}$?

  F $5\frac{2}{7}$

  G $\frac{42}{9}$

  H $\frac{3}{14}$

  I $\frac{7}{28}$

63. Sarah's book bag weighs $18\frac{1}{3}$ pounds. When she removes her math textbook, the book bag weighs $16\frac{3}{8}$ pounds. How much does Sarah's math textbook weigh?

  A $1\frac{23}{24}$ pounds

  B $2\frac{1}{24}$ pounds

  C $2\frac{3}{8}$ pounds

  D $2\frac{23}{24}$ pounds

64. What is the solution to the equation?
$$\frac{4}{9} + \frac{2}{3} = a - \frac{1}{2}$$

  F $\frac{7}{14}$

  G $\frac{11}{18}$

  H $1\frac{1}{9}$

  I $1\frac{11}{18}$

65. Julie needs $2\frac{1}{2}$ times more flour than sugar for a special recipe. If she uses $\frac{3}{4}$ cups of sugar, how much flour does she need for the recipe?

  A $1\frac{7}{8}$ cups

  B 2 cups

  C $2\frac{1}{2}$ cups

  D $3\frac{1}{4}$ cups

66. Mrs. Simon posted the grades for her math class. What is the median of the grades?

  83, 90, 91, 83, 85, 76, 94, 94, 81, 77

  F 83

  G 84

  H 85

  I 96

Holt Middle School Math   Course 1

67. Use the stem-and-leaf plot of last month's daily temperatures to determine how many days the temperature was over 83°.

Stem	Leaves
6	2 4 8 9
7	1 1 3 5 5 6 8 8
8	0 0 1 1 2 2 3 3 6 7
9	2 2 4 7 7 9
10	0 1

A 4 days

B 8 days

C 10 days

D 12 days

68. Find the range of the following set of data.

32, 28, 39, 45, 28, 40, 43, 41

F 17

G 28

H 37

I 39.5

69. Andy recorded the number of points he scored at each basketball game. What is the average number of points that he scored per game?

Games	1	2	3	4	5
Points	8	11	14	7	10

A 8 points

B 9 points

C 10 points

D 11 points

70. Mike recorded the balance of his savings account and organized his data in a stem-and-leaf plot. What was the range of his savings account during the month?

Stem	Leaves
8	2 3 6 8 9
9	0 2 2 5 8 8 9
10	0 3 4 4 5 5 5 9 9
11	2 3 4 5 6 6 7 7 9

8 | 2 means $82

F $37

G $82

H $104

I $119

Holt Middle School Math   Course 1

Diagnostic Test

71. The heights of the boys' basketball team players are listed below in inches. Which measure of central tendency best describes the data?

59 58 55 75 56 53 54

A mean

B median

C mode

D range

72. Janice is a car salesperson. She wants to compare her monthly sales from this year to last year. Which type of graph would best display her data?

F bar graph

G double-line graph

H box-and-whisker plot

I stem-and-leaf plot

73. Use the bar graph below. How many more students like math than English?

Favorite Subject

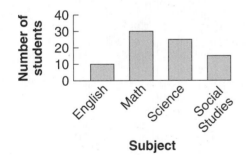

A 5 students

B 10 students

C 15 students

D 20 students

74. Which of the following values is an outlier in this data set?

41, 35, 21, 47, 41, 36, 49, 39

F 21

G 35

H 41

I 49

75. The number of absences per quarter is graphed below. Which statement best describes the graph?

Student Absences

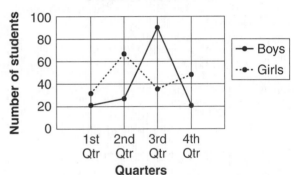

A In the third quarter, more girls were absent than boys.

B In the first two quarters, the boys had better attendance than the girls.

C The girls have the better attendance record.

D Both groups missed about the same number of days in the fourth quarter.

Holt Middle School Math   Course 1

Diagnostic Test

76. Why is this graph misleading?

F Team A is better than Team B.

G The scale does not start at zero.

H The scale is not the same for both teams.

I The graph is not misleading.

77. Which statement is NOT true about rectangles?

A All squares are rectangles.

B All rectangles are quadrilaterals.

C All quadrilaterals are rectangles.

D All rectangles have 90° angles.

78. What types of lines appear in the design?

F skew lines only

G parallel lines only

H intersecting lines only

I intersecting and parallel lines

79. How many lines of symmetry does an isosceles triangle have?

A 0 lines

B 1 lines

C 2 lines

D 3 lines

80. Julie used many shapes to create a design for her school art project. Which of the following pairs of shapes appear to be congruent?

F
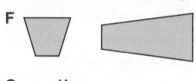

G

H

I

Holt Middle School Math   Course 1

Diagnostic Test

81. Given that lines *m* and *n* are perpendicular, what is the value of *x*?

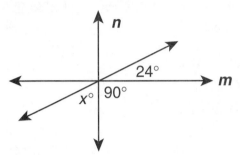

A 24°

B 66°

C 76°

D 90°

82. Which figures represent a slide?

F

G

H

I

83. If a 6 cm by 8 cm rectangle is cut along its diagonal, what two shapes are created?

A two right scalene triangles

B two acute isosceles triangles

C two right equilateral triangles

D two squares

84. Name the figure.

F quadrilateral

G hexagon

H pentagon

I octagon

85. On the figure shown, which line segments are parallel?

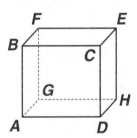

A $\overline{AG}$ and $\overline{DH}$

B $\overline{AB}$ and $\overline{BC}$

C $\overline{BF}$ and $\overline{GH}$

D $\overline{HD}$ and $\overline{EH}$

Holt Middle School Math   Course 1

Diagnostic Test

86. Which of the following shapes can tessellate?

F

G

H

I

87. The angle of a ramp is 15°. What is the supplementary angle?

  A 75°

  B 115°

  C 165°

  D 175°

88. Lou, Lillie, Lucy, and Lyle were practicing free throws. Lou made $\frac{5}{8}$ of his shots. Lillie made $\frac{4}{5}$ of her shots. Lucy made $\frac{3}{4}$ of her shots. Lyle made $\frac{2}{3}$ of her shots. If they each took 120 shots, which player had the best percentage?

  F Lou

  G Lillie

  H Lucy

  I Lyle

89. Four-fifths of Riley's dance team is playing basketball. What percent of her team is NOT playing basketball?

  A $\frac{1}{5}$%

  B 0.40%

  C 20%

  D 80%

90. Which number is equal to $4\frac{7}{100}$?

  F 407%

  G 4.07%

  H 4.7

  I $\frac{4007}{8}$

91. On a sunny day, the flagpole at Jaime's school casts a shadow 40 m long. If Jaime is 1.25 m tall and casts a shadow of 8 m at the same time of day, what is the height of the flagpole?

  A 5.5 m

  B 6.25 m

  C 10 m

  D 32 m

92. Sherry swims 2.5 miles every day in 35 minutes. If she swims at the same rate, how long will it take her to swim 15 miles?

  F 2.25 hours

  G 180 minutes

  H 3.5 hours

  I 210 hours

Diagnostic Test

93. Which pair of rectangles is similar?

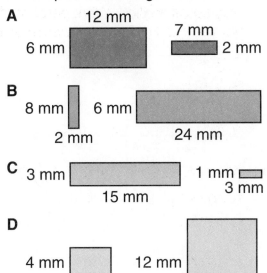

A
12 mm
6 mm
7 mm
2 mm

B
8 mm 6 mm
2 mm 24 mm

C 3 mm
15 mm
1 mm
3 mm

D
4 mm 12 mm
6 mm
10 mm

94. Celia bought a blazer for 25% off. If the original price of the blazer was $75.00, how much did Celia save?

  F $18.75

  G $25.00

  H $42.00

  I $56.25

95. What is 60% written as a fraction in simplest form?

  A $\frac{6}{10}$

  B $\frac{60}{1000}$

  C $\frac{3}{5}$

  D $\frac{1}{2}$

96. The scale factor for a model plane is 2 in. = 25 ft. If the length of the model is 8 inches, how long is the plane?

  F 100 inches

  G 8 ft

  H 80 ft

  I 100 ft

97. What is 15% of 107?

  A 1.605

  B 16.05

  C 160.5

  D 1,605

98. John is copying files from his computer. He has copied 75% of his files in 45 minutes. How long will it take him to copy the rest of his files?

  F 15 minutes

  G 30 minutes

  H 60 minutes

  I 90 minutes

99. The cheerleading team is sponsoring a car wash. The team can wash 8 cars in two hours. How many cars can the team wash in 6 hours?

  A 8 cars

  B 14 cars

  C 16 cars

  D 24 cars

Holt Middle School Math   Course 1

Diagnostic Test

100. How many 8 inch long ribbons can Mona make from a 4 foot piece of material?

   F 6 ribbons

   G 8 ribbons

   H 10 ribbons

   I 12 ribbons

101. Anna's lunch bill came to $25.50 before the tip. She tipped the server $4.59. What percent of the bill was the tip?

   A 15%

   B 18%

   C 21%

   D 24%

102. Find the coordinates of point *J*.

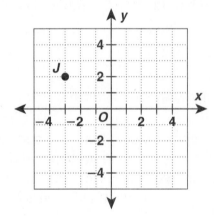

   F (2, −3)

   G (−2, −3)

   H (3, −2)

   I (−3, 2)

103. Which point is plotted at (−3, −3) on the unit grid?

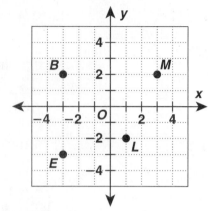

   A point *B*

   B point *M*

   C point *E*

   D point *L*

104. Which of the following is NOT an integer?

   F −2

   G $-\frac{1}{2}$

   H 0

   I 10

105. During a football game the offense had a gain of 15 yards, a loss of 4 yards, and a loss of 20 yards before they had to punt. What was the net gain/loss of this offensive drive?

   A −4 yards

   B −9 yards

   C 15 yards

   D 20 yards

Holt Middle School Math   **Course 1**

Diagnostic Test

106. What is the distance between point *A* and point *B* on the number line?

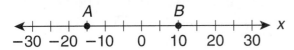

F −25

G −15

H 15

I 25

107. What is the sum of −234 and 435?

A −669

B −192

C 201

D 669

108. The temperature at sunrise was −2°F. By noon the temperature had increased by 19°F. What was the temperature at noon?

F 17°F

G 21°F

H 23°F

I 25°F

109. Ashley was fossil hunting when she found a fossil 8 feet below sea level. She then climbed 18 feet up and found another fossil. How many feet above sea level was the second fossil?

A 8 feet

B 10 feet

C 20 feet

D 24 feet

110. Bill lost $175 a month on the stock market over the past 9 months. What was the overall change in the value of his stock?

F −$19.44

G −$184.00

H −$1,557.00

I −$1,575.00

111. What is the value of $m \times (-3)$ when $m = -7$?

A −21

B −10

C −4

D 21

112. What is the solution to the following equation?

$x - 7 = -16$

F $x = -23$

G $x = -9$

H $x = 9$

I $x = -112$

113. What two integers have a product of 64 and a sum of −20?

A −8 and −8

B −16 and −4

C −16 and 4

D −12 and −8

Holt Middle School Math   Course 1

Diagnostic Test

114. Tommy went deep sea diving. He recorded the depth of his dives in the table below. What was the average depth of his dives?

Dive 1	Dive 2	Dive 3	Dive 4	Dive 5
−25 ft	−14 ft	−32 ft	−22 ft	−37 ft

    F −28 ft   **H** 26 ft

    G −26 ft   **I** 28 ft

115. A submarine is 450 meters below sea level. An airplane is 832 meters above sea level. What is the total distance between the plane and the submarine?

    A −382 m   **C** 641 m

    B 382 m   **D** 1,282 m

116. Which of the following is a triangular prism?

F

G

H

I

117. A triangle has a 24 cm base and a height of 15 cm. If the height is reduced by 5, what will be the new area of the triangle?

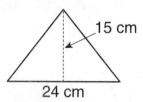

    A 120 cm^2

    B 180 cm^2

    C 240 cm^2

    D 360 cm^2

118. What is the circumference of a circle with a radius of 5 mm? Use 3.14 for π.

    F 15.7 mm

    G 25.4 mm

    H 31.4 mm

    I 62.8 mm

Diagnostic Test

119. Glen must break the box down so it is flat. Which shape matches the box once it has been broken down?

A

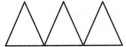

B

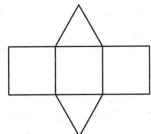

C

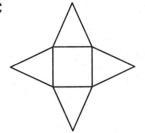

D

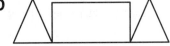

120. Maria has a fence around her pool yard. Each fence post is 12 ft apart. What is the perimeter of Maria's pool yard?

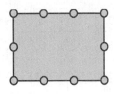

F 10 ft
G 60 ft
H 80 ft
I 120 ft

121. A round area rug has a radius of three feet. How much area does the rug cover? Use 3.14 for π.
A 9.42 ft^2
B 18.85 ft^2
C 28.26 ft^2
D 56.55 ft^2

122. Danielle has 1,000 in^2 of wrapping paper. How many of the following gift boxes can she wrap?

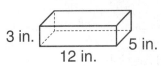

3 in.   12 in.   5 in.

F 2 boxes
G 3 boxes
H 4 boxes
I 5 boxes

Holt Middle School Math   Course 1

Name _____ Date _____ Class _____

Diagnostic Test

123. Roberta is buying new carpet for a room that measures 22 feet × 15 feet. If the carpet costs $8.00 per square foot, how much will the new carpet cost?

A $330.00

B $880.00

C $1,220.00

D $2,640.00

124. A cylinder has a height of 4 cm and a diameter of 3 cm. What is the volume of the bucket? Use 3.14 for π.

F 12 cm^3

G 21.98 cm^3

H 28.26 cm^3

I 113.14 cm^3

125. What is the perimeter of the shaded area?

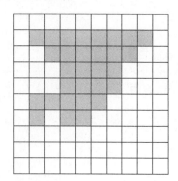

A 28 units

B 30 units

C 34 units

D 32 units

126. Which of the following cylinders has the greatest volume?

F   10 cm

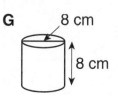
15 cm

G   8 cm

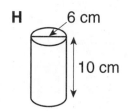
8 cm

H   6 cm

10 cm

I   12 cm

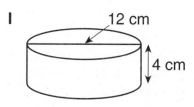
4 cm

127. Ryan flipped a number cube 75 times and recorded the results below.

Number on Cube	1	2	3	4	5	6
Frequency	12	10	14	14	12	13

What is the difference between the theoretical probability that the number cube will land on a 4 and the experimental probability that the number cube will land on a 4?

A 2%   **C** 18%

B 16%   **D** 34%

Holt Middle School Math   Course 1

Diagnostic Test

128. Cathy randomly pulls one block from the box. What is the probability that it will be black?

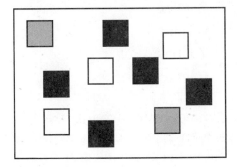

F $\dfrac{2}{5}$

G $\dfrac{1}{2}$

H $\dfrac{2}{3}$

I $\dfrac{4}{5}$

129. Mia has 4 shirts, 2 pairs of shoes, and 4 pairs of pants. What is the number of different possible outfits that she could wear?

A 8 outfits

B 10 outfits

C 16 outfits

D 32 outfits

130. The code to Ms. Kyle's alarm system is 5 digits. Each digit can be a number from 1 to 5, but each number can only be used once. How many different combinations can Mr. Kaiser come up with for his pass code?

F 5

G 20

H 100

I 120

131. Lisa randomly pulled one marble out of the sack. What is the probability that the marble is NOT white?

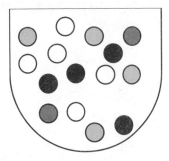

A $\dfrac{1}{5}$

B $\dfrac{1}{3}$

C $\dfrac{1}{2}$

D $\dfrac{2}{3}$

132. Fred is buying a coat. The coat is sold in three colors: red, black, brown, may come with a zipper or buttons, and may have a hood. What is the probability that Fred will randomly buy a black, zippered coat with no hood?

F $\dfrac{1}{12}$

G $\dfrac{1}{6}$

H $\dfrac{3}{12}$

I $\dfrac{2}{3}$

Holt Middle School Math   Course 1

Name _____ Date _____ Class _____

Diagnostic Test

133. Sam has two fair number cubes. He needs to roll a 6 to win the board game. What is the probability that he will roll a sum of six on the number cubes?

A $\frac{1}{12}$

B $\frac{5}{36}$

C $\frac{1}{6}$

D $\frac{2}{3}$

134. Jack can pick one drink and one piece of fruit for his snack. The drink choices are milk, fruit juice, or tea. The fruit choices are an apple, a pear, a banana, or a kiwi. How many different snack combinations are there?

F 6

G 7

H 9

I 12

135. The probability of a fair coin landing on heads is $\frac{1}{2}$. What is the probability of the coin landing on heads three consecutive times?

A $\frac{1}{2}$

B $\frac{1}{4}$

C $\frac{1}{8}$

D $\frac{3}{8}$

136. If you roll a fair number cube 42 times, how many times would you expect to roll a 2?

F 2 times

G 6 times

H 7 times

I 8 times

137. Kim needs to take math, art, science, and English. How many different schedules can Kim make?

A 4

B 8

C 24

D 64

138. If you spin a spinner with 3 possible outcomes 3 times, how many possible outcomes are there?

F 3 outcomes

G 9 outcomes

H 18 outcomes

I 27 outcomes

139. What is the missing value in the table given that $y = 5x - 12$?

x	y
3	3
1	−7
−1	?

A −17

B −12

C 7

D 18

Diagnostic Test

140. The population in Carla's town is three times smaller than the population in Tonya's town. Which equation correctly represents the population if y represents the population in Carla's town and x represents the population in Tonya's town?

F $y = 3x$

G $y = 3x^2$

H $y = \frac{1}{3}x$

I $y = x - 3$

141. Ryan has 75 minutes to work on his science and math homework. Which equation correctly relates x, the time Ryan has to complete his science homework, and y, the time he has to complete his math homework?

A $x - 75 = y$

B $x - y = 75$

C $\frac{x}{y} = 75$

D $x + y = 75$

142. In the equation $y = -7x + 3$, what is the value of y when $x = -1$?

F -4

G 3

H 10

I 11

143. Shannon was graphing equations of lines on a coordinate plane. Which is the equation of a line that would have contained the point $(5, -2)$?

A $y = x - 2$

B $y = x - 7$

C $y = 2x - 14$

D $y = -\frac{x}{2} + 2$

144. Look at the input-output table. What is the rule for this table?

Input	Output
4	9
6	13
8	17

F $y = 2x + 1$

G $y = -2x + 18$

H $y = 3x - 5$

I $y = x + 11$

145. The vertices of a triangle are $(3, 1)$, $(4, 9)$, and $(8, -1)$. What are the coordinates of the triangle after a translation of 5 units to the left?

A $(3, -4)$ $(4, 4)$, $(8, -6)$

B $(-2, -4)$ $(-1, 4)$, $(3, -6)$

C $(-2, 1)$, $(-1, 9)$, $(3, -1)$

D $(8, -1)$ $(9, 9)$, $(13, -1)$

Holt Middle School Math   Course 1

Diagnostic Test

146. Which ordered pair is a solution for the equation?

$y = -3x + 7$

F $(0, -7)$

G $(-2, 13)$

H $(7, 0)$

I $(1, 10)$

147. What is the coordinate of B' after a refection across the y axis?

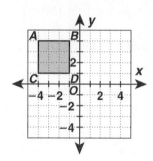

A $(-1, 3)$

B $(4, 4)$

C $(1, 4)$

D $(1, 1)$

148. Which ordered pair is NOT a solution for the equation $y = \frac{1}{2}x - 6$?

F $(0, -6)$

G $(2, -5)$

H $(4, -2)$

I $(-2, -7)$

149. Which graph corresponds to the equation?

$y = x + 2$

A

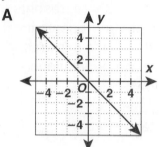

B

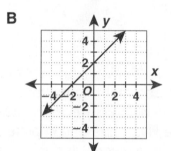

C

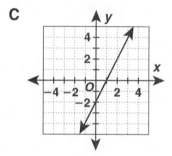

D

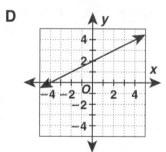

Holt Middle School Math   Course 1

Diagnostic Test Answer Sheet

1. (A) (B) (C) (D) See Lesson 1-3.
2. (F) (G) (H) (I) See Lesson 1-6.
3. (A) (B) (C) (D) See Lesson 1-6.
4. (F) (G) (H) (I) See Lesson 1-2.
5. (A) (B) (C) (D) See Lesson 1-4.
6. (F) (G) (H) (I) See Lesson 1-7.
7. (A) (B) (C) (D) See Lesson 1-2.
8. (F) (G) (H) (I) See Lesson 1-7.
9. (A) (B) (C) (D) See Lesson 1-4.
10. (F) (G) (H) (I) See Lesson 1-7.
11. (A) (B) (C) (D) See Lesson 1-1.
12. (F) (G) (H) (I) See Lesson 1-3.
13. (A) (B) (C) (D) See Lesson 1-2.
14. (F) (G) (H) (I) See Lesson 2-2.
15. (A) (B) (C) (D) See Lesson 2-1.
16. (F) (G) (H) (I) See Lesson 2-1.
17. (A) (B) (C) (D) See Lesson 2-3.
18. (F) (G) (H) (I) See Lesson 2-6.
19. (A) (B) (C) (D) See Lesson 2-1.
20. (F) (G) (H) (I) See Lesson 2-4.
21. (A) (B) (C) (D) See Lesson 2-7.
22. (F) (G) (H) (I) See Lesson 2-3.
23. (A) (B) (C) (D) See Lesson 2-6.
24. (F) (G) (H) (I) See Lesson 2-3.
25. (A) (B) (C) (D) See Lesson 2-1.
26. (F) (G) (H) (I) See Lesson 2-6.
27. (A) (B) (C) (D) See Lesson 2-1.
28. (F) (G) (H) (I) See Lesson 3-6.
29. (A) (B) (C) (D) See Lesson 3-1.

30. (F) (G) (H) (I) See Lesson 3-5.
31. (A) (B) (C) (D) See Lesson 3-4.
32. (F) (G) (H) (I) See Lesson 3-7.
33. (A) (B) (C) (D) See Lesson 3-9.
34. (F) (G) (H) (I) See Lesson 3-1.
35. (A) (B) (C) (D) See Lesson 3-4.
36. (F) (G) (H) (I) See Lesson 3-10.
37. (A) (B) (C) (D) See Lesson 3-1.
38. (F) (G) (H) (I) See Lesson 3-3.
39. (A) (B) (C) (D) See Lesson 3-4.
40. (F) (G) (H) (I) See Lesson 3-2.
41. (A) (B) (C) (D) See Lesson 4-3.
42. (F) (G) (H) (I) See Lesson 4-1.
43. (A) (B) (C) (D) See Lesson 4-2.
44. (F) (G) (H) (I) See Lesson 4-5.
45. (A) (B) (C) (D) See Lesson 4-2.
46. (F) (G) (H) (I) See Lesson 4-4.
47. (A) (B) (C) (D) See Lesson 4-6.
48. (F) (G) (H) (I) See Lesson 4-8.
49. (A) (B) (C) (D) See Lesson 4-9.
50. (F) (G) (H) (I) See Lesson 4-8.
51. (A) (B) (C) (D) See Lesson 4-9.
52. (F) (G) (H) (I) See Lesson 4-4.
53. (A) (B) (C) (D) See Lesson 4-5.
54. (F) (G) (H) (I) See Lesson 5-1.
55. (A) (B) (C) (D) See Lesson 5-2.
56. (F) (G) (H) (I) See Lesson 5-3.
57. (A) (B) (C) (D) See Lesson 5-7.
58. (F) (G) (H) (I) See Lesson 5-7.

Holt Middle School Math Course 1

Name _____ Date _____ Class _____

Diagnostic Test Answer Sheet

59. (A) (B) (C) (D)   See Lesson 5-5.
60. (F) (G) (H) (I)   See Lesson 5-10.
61. (A) (B) (C) (D)   See Lesson 5-3.
62. (F) (G) (H) (I)   See Lesson 5-3.
63. (A) (B) (C) (D)   See Lesson 5-10.
64. (F) (G) (H) (I)   See Lesson 5-10.
65. (A) (B) (C) (D)   See Lesson 6-2.
66. (F) (G) (H) (I)   See Lesson 6-2.
67. (A) (B) (C) (D)   See Lesson 6-9.
68. (F) (G) (H) (I)   See Lesson 6-2.
69. (A) (B) (C) (D)   See Lesson 6-2.
70. (F) (G) (H) (I)   See Lesson 6-9.
71. (A) (B) (C) (D)   See Lesson 6-2.
72. (F) (G) (H) (I)   See Lesson 6-7.
73. (A) (B) (C) (D)   See Lesson 6-4.
74. (F) (G) (H) (I)   See Lesson 6-3.
75. (A) (B) (C) (D)   See Lesson 6-7.
76. (F) (G) (H) (I)   See Lesson 6-8.
77. (A) (B) (C) (D)   See Lesson 7-6.
78. (F) (G) (H) (I)   See Lesson 7-4.
79. (A) (B) (C) (D)   See Lesson 7-11.
80. (F) (G) (H) (I)   See Lesson 7-9.
81. (A) (B) (C) (D)   See Lesson 7-3.
82. (F) (G) (H) (I)   See Lesson 7-10.
83. (A) (B) (C) (D)   See Lesson 7-6.
84. (F) (G) (H) (I)   See Lesson 7-7.
85. (A) (B) (C) (D)   See Lesson 7-4.
86. (F) (G) (H) (I)   See Lesson 7-12.
87. (A) (B) (C) (D)   See Lesson 7-4.

88. (F) (G) (H) (I)   See Lesson 8-9.
89. (A) (B) (C) (D)   See Lesson 8-8.
90. (F) (G) (H) (I)   See Lesson 8-8.
91. (A) (B) (C) (D)   See Lesson 8-5.
92. (F) (G) (H) (I)   See Lesson 8-1.
93. (A) (B) (C) (D)   See Lesson 8-4.
94. (F) (G) (H) (I)   See Lesson 8-10.
95. (A) (B) (C) (D)   See Lesson 8-8.
96. (F) (G) (H) (I)   See Lesson 8-6.
97. (A) (B) (C) (D)   See Lesson 8-10.
98. (F) (G) (H) (I)   See Lesson 8-1.
99. (A) (B) (C) (D)   See Lesson 8-2.
100. (F) (G) (H) (I)   See Lesson 8-3.
101. (A) (B) (C) (D)   See Lesson 8-10.
102. (F) (G) (H) (I)   See Lesson 9-3.
103. (A) (B) (C) (D)   See Lesson 9-3.
104. (F) (G) (H) (I)   See Lesson 9-1.
105. (A) (B) (C) (D)   See Lesson 9-4.
106. (F) (G) (H) (I)   See Lesson 9-1.
107. (A) (B) (C) (D)   See Lesson 9-4.
108. (F) (G) (H) (I)   See Lesson 9-4.
109. (A) (B) (C) (D)   See Lesson 9-4.
110. (F) (G) (H) (I)   See Lesson 9-7.
111. (A) (B) (C) (D)   See Lesson 9-6.
112. (F) (G) (H) (I)   See Lesson 9-8.
113. (A) (B) (C) (D)   See Lesson 9-6.
114. (F) (G) (H) (I)   See Lesson 9-7.
115. (A) (B) (C) (D)   See Lesson 9-4.
116. (F) (G) (H) (I)   See Lesson 10-6.

Holt Middle School Math Course 1

Diagnostic Test Answer Sheet

117. (A) (B) (C) (D) See Lesson 10-2.

118. (F) (G) (H) (I) See Lesson 10-5.

119. (A) (B) (C) (D) See Lesson 10-6.

120. (F) (G) (H) (I) See Lesson 10-1.

121. (A) (B) (C) (D) See Lesson 10-5.

122. (F) (G) (H) (I) See Lesson 10-7.

123. (A) (B) (C) (D) See Lesson 10-2.

124. (F) (G) (H) (I) See Lesson 10-8.

125. (A) (B) (C) (D) See Lesson 10-1.

126. (F) (G) (H) (I) See Lesson 10-9.

127. (A) (B) (C) (D) See Lesson 11-3.

128. (F) (G) (H) (I) See Lesson 11-1.

129. (A) (B) (C) (D) See Lesson 11-5.

130. (F) (G) (H) (I) See Lesson 11-5.

131. (A) (B) (C) (D) See Lesson 11-1.

132. (F) (G) (H) (I) See Lesson 11-4.

133. (A) (B) (C) (D) See Lesson 11-5.

134. (F) (G) (H) (I) See Lesson 11-4.

135. (A) (B) (C) (D) See Lesson 11-5.

136. (F) (G) (H) (I) See Lesson 11-2.

137. (A) (B) (C) (D) See Lesson 11-4.

138. (F) (G) (H) (I) See Lesson 11-4.

139. (A) (B) (C) (D) See Lesson 12-1.

140. (F) (G) (H) (I) See Lesson 12-1.

141. (A) (B) (C) (D) See Lesson 12-1.

142. (F) (G) (H) (I) See Lesson 12-1.

143. (A) (B) (C) (D) See Lesson 12-2.

144. (F) (G) (H) (I) See Lesson 12-1.

145. (A) (B) (C) (D) See Lesson 12-3.

146. (F) (G) (H) (I) See Lesson 12-1.

147. (A) (B) (C) (D) See Lesson 12-4.

148. (F) (G) (H) (I) See Lesson 12-1.

149. (A) (B) (C) (D) See Lesson 12-2.

Holt Middle School Math Course 1

Diagnostic Test Answer Sheet

#	A/F	B/G	C/H	D/I	Lesson
1.	(A)	(B)	(C)	●	See Lesson 1-3.
2.	(F)	(G)	(H)	●	See Lesson 1-6.
3.	(A)	(B)	(C)	●	See Lesson 1-6.
4.	(F)	(G)	●	(I)	See Lesson 1-2.
5.	(A)	●	(C)	(D)	See Lesson 1-4.
6.	●	(G)	(H)	(I)	See Lesson 1-7.
7.	(A)	(B)	●	(D)	See Lesson 1-2.
8.	(F)	(G)	●	(I)	See Lesson 1-7.
9.	(A)	(B)	●	(D)	See Lesson 1-4.
10.	(F)	(G)	●	(I)	See Lesson 1-7.
11.	(A)	●	(C)	(D)	See Lesson 1-1.
12.	(F)	●	(H)	(I)	See Lesson 1-3.
13.	(A)	(B)	●	(D)	See Lesson 1-2.
14.	(F)	(G)	●	(I)	See Lesson 2-2.
15.	(A)	(B)	(C)	●	See Lesson 2-1.
16.	(F)	●	(H)	(I)	See Lesson 2-1.
17.	(A)	(B)	(C)	●	See Lesson 2-3.
18.	(F)	(G)	●	(I)	See Lesson 2-6.
19.	(A)	●	(C)	(D)	See Lesson 2-1.
20.	●	(G)	(H)	(I)	See Lesson 2-4.
21.	(A)	(B)	(C)	●	See Lesson 2-7.
22.	(F)	●	(H)	(I)	See Lesson 2-3.
23.	(A)	(B)	(C)	●	See Lesson 2-6.
24.	(F)	●	(H)	(I)	See Lesson 2-3.
25.	(A)	●	(C)	(D)	See Lesson 2-1.
26.	(F)	(G)	●	(I)	See Lesson 2-6.
27.	(A)	●	(C)	(D)	See Lesson 2-1.
28.	(F)	●	(H)	(I)	See Lesson 3-6.
29.	●	(B)	(C)	(D)	See Lesson 3-1.
30.	(F)	(G)	●	(I)	See Lesson 3-5.
31.	(A)	(B)	(C)	●	See Lesson 3-4.
32.	(F)	(G)	●	(I)	See Lesson 3-7.
33.	(A)	(B)	(C)	●	See Lesson 3-9.
34.	(F)	●	(H)	(I)	See Lesson 3-1.
35.	(A)	●	(C)	(D)	See Lesson 3-4.
36.	●	(G)	(H)	(I)	See Lesson 3-10.
37.	●	(B)	(C)	(D)	See Lesson 3-1.
38.	(F)	●	(H)	(I)	See Lesson 3-3.
39.	(A)	(B)	(C)	●	See Lesson 3-4.
40.	(F)	(G)	●	(I)	See Lesson 3-2.
41.	(A)	(B)	(C)	●	See Lesson 4-3.
42.	(F)	●	(H)	(I)	See Lesson 4-1.
43.	(A)	●	(C)	(D)	See Lesson 4-2.
44.	(F)	●	(H)	(I)	See Lesson 4-5.
45.	(A)	●	(C)	(D)	See Lesson 4-2.
46.	(F)	(G)	●	(I)	See Lesson 4-4.
47.	(A)	(B)	(C)	●	See Lesson 4-6.
48.	(F)	(G)	●	(I)	See Lesson 4-8.
49.	(A)	(B)	●	(D)	See Lesson 4-9.
50.	(F)	●	(H)	(I)	See Lesson 4-8.
51.	(A)	(B)	(C)	●	See Lesson 4-9.
52.	(F)	(G)	●	(I)	See Lesson 4-4.
53.	(A)	●	(C)	(D)	See Lesson 4-5.
54.	(F)	●	(H)	(I)	See Lesson 5-1.
55.	(A)	(B)	●	(D)	See Lesson 5-2.
56.	(F)	●	(H)	(I)	See Lesson 5-3.
57.	(A)	(B)	●	(D)	See Lesson 5-7.
58.	(F)	(G)	●	(I)	See Lesson 5-7.

Diagnostic Test Answer Sheet

59. Ⓐ Ⓑ ● Ⓓ   See Lesson 5-5.
60. Ⓕ Ⓖ ● Ⓘ   See Lesson 5-10.
61. Ⓐ Ⓑ Ⓒ ●   See Lesson 5-3.
62. Ⓕ Ⓖ ● Ⓘ   See Lesson 5-3.
63. ● Ⓑ Ⓒ Ⓓ   See Lesson 5-10.
64. Ⓕ Ⓖ Ⓗ ●   See Lesson 5-10.
65. ● Ⓑ Ⓒ Ⓓ   See Lesson 6-2.
66. Ⓕ ● Ⓗ Ⓘ   See Lesson 6-2.
67. Ⓐ Ⓑ ● Ⓓ   See Lesson 6-9.
68. ● Ⓖ Ⓗ Ⓘ   See Lesson 6-2.
69. Ⓐ Ⓑ ● Ⓓ   See Lesson 6-2.
70. ● Ⓖ Ⓗ Ⓘ   See Lesson 6-9.
71. Ⓐ ● Ⓒ Ⓓ   See Lesson 6-2.
72. Ⓕ ● Ⓗ Ⓘ   See Lesson 6-7.
73. Ⓐ Ⓑ Ⓒ ●   See Lesson 6-4.
74. ● Ⓖ Ⓗ Ⓘ   See Lesson 6-3.
75. Ⓐ ● Ⓒ Ⓓ   See Lesson 6-7.
76. Ⓕ ● Ⓗ Ⓘ   See Lesson 6-8.
77. Ⓐ Ⓑ ● Ⓓ   See Lesson 7-6.
78. Ⓕ Ⓖ Ⓗ ●   See Lesson 7-4.
79. Ⓐ ● Ⓒ Ⓓ   See Lesson 7-11.
80. Ⓕ Ⓖ Ⓗ ●   See Lesson 7-9.
81. Ⓐ ● Ⓒ Ⓓ   See Lesson 7-3.
82. Ⓕ Ⓖ ● Ⓘ   See Lesson 7-10.
83. ● Ⓑ Ⓒ Ⓓ   See Lesson 7-6.
84. Ⓕ Ⓖ ● Ⓘ   See Lesson 7-7.
85. ● Ⓑ Ⓒ Ⓓ   See Lesson 7-4.
86. Ⓕ Ⓖ Ⓗ ●   See Lesson 7-12.
87. Ⓐ Ⓑ ● Ⓓ   See Lesson 7-4.

88. Ⓕ ● Ⓗ Ⓘ   See Lesson 8-9.
89. Ⓐ Ⓑ ● Ⓓ   See Lesson 8-8.
90. ● Ⓖ Ⓗ Ⓘ   See Lesson 8-8.
91. Ⓐ ● Ⓒ Ⓓ   See Lesson 8-5.
92. Ⓕ Ⓖ ● Ⓘ   See Lesson 8-1.
93. Ⓐ ● Ⓒ Ⓓ   See Lesson 8-4.
94. ● Ⓖ Ⓗ Ⓘ   See Lesson 8-10.
95. Ⓐ Ⓑ ● Ⓓ   See Lesson 8-8.
96. Ⓕ Ⓖ Ⓗ ●   See Lesson 8-6.
97. Ⓐ ● Ⓒ Ⓓ   See Lesson 8-10.
98. ● Ⓖ Ⓗ Ⓘ   See Lesson 8-1.
99. Ⓐ Ⓑ Ⓒ ●   See Lesson 8-2.
100. ● Ⓖ Ⓗ Ⓘ   See Lesson 8-3.
101. Ⓐ ● Ⓒ Ⓓ   See Lesson 8-10.
102. Ⓕ Ⓖ Ⓗ ●   See Lesson 9-3.
103. Ⓐ Ⓑ ● Ⓓ   See Lesson 9-3.
104. Ⓕ ● Ⓗ Ⓘ   See Lesson 9-1.
105. Ⓐ ● Ⓒ Ⓓ   See Lesson 9-4.
106. Ⓕ Ⓖ Ⓗ ●   See Lesson 9-1.
107. Ⓐ Ⓑ ● Ⓓ   See Lesson 9-4.
108. ● Ⓖ Ⓗ Ⓘ   See Lesson 9-4.
109. Ⓐ ● Ⓒ Ⓓ   See Lesson 9-4.
110. Ⓕ Ⓖ Ⓗ ●   See Lesson 9-7.
111. Ⓐ Ⓑ Ⓒ ●   See Lesson 9-6.
112. Ⓕ ● Ⓗ Ⓘ   See Lesson 9-8.
113. Ⓐ ● Ⓒ Ⓓ   See Lesson 9-6.
114. Ⓕ ● Ⓗ Ⓘ   See Lesson 9-7.
115. Ⓐ Ⓑ Ⓒ ●   See Lesson 9-4.
116. Ⓕ Ⓖ ● Ⓘ   See Lesson 10-6.

27B

Holt Middle School Math Course 1

Diagnostic Test Answer Sheet

117. ● (B) (C) (D)   See Lesson 10-2.
118. (F) (G) ● (I)   See Lesson 10-5.
119. (A) ● (C) (D)   See Lesson 10-6.
120. (F) (G) (H) ●   See Lesson 10-1.
121. (A) (B) ● (D)   See Lesson 10-5.
122. (F) (G) ● (I)   See Lesson 10-7.
123. (A) (B) (C) ●   See Lesson 10-2.
124. (F) (G) ● (I)   See Lesson 10-8.
125. (A) (B) ● (D)   See Lesson 10-1.
126. ● (G) (H) (I)   See Lesson 10-9.
127. ● (B) (C) (D)   See Lesson 11-3.
128. (F) ● (H) (I)   See Lesson 11-1.
129. (A) (B) (C) ●   See Lesson 11-5.
130. (F) (G) (H) ●   See Lesson 11-5.
131. (A) (B) (C) ●   See Lesson 11-1.
132. ● (G) (H) (I)   See Lesson 11-4.
133. (A) ● (C) (D)   See Lesson 11-5.

134. (F) (G) (H) ●   See Lesson 11-4.
135. (A) (B) ● (D)   See Lesson 11-5.
136. (F) (G) ● (I)   See Lesson 11-2.
137. (A) (B) ● (D)   See Lesson 11-4.
138. (F) (G) (H) ●   See Lesson 11-4.
139. ● (B) (C) (D)   See Lesson 12-1.
140. (F) (G) ● (I)   See Lesson 12-1.
141. (A) (B) (C) ●   See Lesson 12-1.
142. (F) (G) ● (I)   See Lesson 12-1.
143. (A) ● (C) (D)   See Lesson 12-2.
144. ● (G) (H) (I)   See Lesson 12-1.
145. (A) (B) ● (D)   See Lesson 12-3.
146. (F) ● (H) (I)   See Lesson 12-1.
147. (A) (B) ● (D)   See Lesson 12-4.
148. (F) (G) ● (I)   See Lesson 12-1.
149. (A) ● (C) (D)   See Lesson 12-2.

Holt Middle School Math Course 1

Name _____ Date _____ Class _____

Test Taking Strategy
Multiple Choice: Find a Pattern

Chapter 1

You can sometimes find the answer to a multiple choice question by using the answer choices to work backwards.

Example 1 What is the missing number in the sequence?

2   8   18   ___   50   72

A 28   **B** 32   **C** 40   **D** 46

Solution: Look for a pattern to find how the numbers given in the sequence were found. Look for something in common with the numbers.

$1 \cdot 2 = 2$
$4 \cdot 2 = 8$
$9 \cdot 2 = 18$ Each number in the sequence is a multiple of 2.
$25 \cdot 2 = 50$
$36 \cdot 2 = 72$

Notice that the factor multiplied by 2 is a perfect square.
$1^2 = 1, 2^2 = 4, 3^2 = 9, 5^2 = 25$, and $6^2 = 36$.
So, $4^2 = 16$, and $16 \times 2 = 32$. The missing number in the sequence is 32. The correct answer is choice B.

Answer choices to multiple choice questions usually contain distracters. Distracters are values that are arrived at by making a common misjudgment or a simple error in a calculation. You need to double check your calculation and reread the question statement in order to avoid choosing one of the given distracters as an answer choice.

Example 2 Use the pattern to find the units digit of 3^{10}.

$3^1 = \underline{3}$ $3^2 = \underline{9}$ $3^3 = 2\underline{7}$ $3^4 = 8\underline{1}$ $3^5 = 24\underline{3}$

F 3   **G** 9   **H** 7   **I** 1

Solution: Once you notice that the sequence of numbers is 3, 9, 7, 1 you can continue the pattern to find the units digit of 3^{10}. 3^6 has a $\underline{9}$ in the units digit. 3^7 has a $\underline{7}$ in the units digit. 3^8 has a $\underline{1}$ in the units digit. 3^9 has a $\underline{3}$ in the units digit. 3^{10} has a $\underline{9}$ in the units digit. The correct answer is choice G.

Notice that the other answer choices contain the distracters 3, 7, and 1. If you had made a simple error in your reasoning, you might have selected the wrong answer choice.

Holt Middle School Math   Course 1

Name _____ Date _____ Class _____

Test Taking Strategy
Chapter 1, continued

Exercises Possible answers are given.

Multiple Choice Which value is the missing number in the sequence?

46   42   38   ___   30   26

A 36   **B** 35   **C** 34   **D** 33

1. Explain how you can work backwards to determine the missing value.

   I can substitute each given answer choice into the pattern and see if that value is part of the sequence.

2. What is the pattern and which value completes the sequence?

   Each value is four less than the previous value. The missing number is 34, Choice C.

Multiple Choice Use the following pattern to find the units digit of 4^{12}.

$4^1 = \underline{4}$ $4^2 = 1\underline{6}$ $4^3 = 6\underline{4}$ $4^4 = 25\underline{6}$ $4^5 = 1,02\underline{4}$

F 2   **G** 4   **H** 6   **I** 8

3. Which answer choices are NOT considered to be distracters? Why?

   Choice F and Choice I; The pattern only includes the numbers 4 and 6 and the choices given in F and I cannot be arrived at by a simple mistake or misjudgment.

4. The correct answer is 6, Choice H. Explain why Choice G is a distracter.

   Choice G is a distracter because you could possible make a minor error in calculating the 12 value of the sequence, resulting in the answer of 4.

Holt Middle School Math   Course 1

Test Taking Strategy
Chapter 2

Extended Response questions are scored using a scoring rubric. You are expected to show all your work in detail. Use complete sentences to explain your thought process. Below is a scoring rubric that is used to score the answers to the following exercise.

Scoring Rubric

- **4 points:** Student writes and correctly solves an equation. Student defines the variable. Student answers the question with clear and complete explanations.

- **3 points:** Student writes and correctly solves an equation but does not show all steps or does not give a clear explanation.

- **2 points:** Student gives correct answer but does not write and solve an equation, or student writes and solves incorrect equation.

- **1 point:** Student writes correct equation but does not solve the equation or give any explanation.

- **0 points:** Student gives incorrect solution, no equation or explanation, or no response.

Example Lisa has a collection of mystery books. She cleans her bookshelf and gives Cindy 18 books. She now has 23 books on her shelf. Write an equation to determine how many books Lisa originally had on her shelf, and then solve the equation. Show your work.

The following is an example of a 4-point response.
Let $b =$ the original number of books on Lisa's shelf.

$$b - 18 = 23$$
$$b - 18 + 18 = 23 + 18 \qquad \text{Write and solve an equation.}$$
$$b = 41$$

Lisa originally had 41 books on her shelf.

The following is an example of a 3-point response.

$$b - 18 = 23$$
$$b - 18 + 18 = 23 + 18 \qquad \text{Write and solve an equation.}$$
$$b = 41$$

This is only a 2-point response because the student didn't define the variable or write the answer in a complete sentence.

Name _____ Date _____ Class _____

Test Taking Strategy
Chapter 2, continued

Exercises

Use the scoring rubric for each question.

Scoring Rubric

- **4 points:** Student writes and correctly solves an equation. Student defines the variable. Student answers the question with clear and complete explanations.

- **3 points:** Student writes and correctly solves an equation but does not show all steps or does not give a clear explanation.

- **2 points:** Student gives correct answer but does not write and solve an equation, or student writes and solves incorrect equation.

- **1 point:** Student writes correct equation but does not solve the equation or give any explanation.

- **0 points:** Student gives incorrect solution, no equation or explanation, or no response.

---

1. Jeremy collects baseball cards. After selling 21 cards to his friend Mike, he has 105 cards remaining in his collection. Write an equation to determine how many cards Jeremy originally had in his collection, and then solve the equation. Show your work.

   3-point response:
   Let c = the original number of cards in the collection.
   $$c - 21 = 105$$
   $$c = 126$$

   Explain why the student received only 3 points for this answer.

   Possible answer: The student failed to show each step of the work. The student also failed to provide reasoning or write the answer in a complete sentence.

2. To receive all the points available on an answer to an extended response question, what must your answer always include?

   A a diagram or a sketch
   B formulas
   C an explanation or reasoning process
   D a visual data display

Test Taking Strategy
Chapter 3

Eliminating answer choices that you know are incorrect or
unreasonable is an excellent test taking strategy. Use mental math
and estimation techniques to help you decide which answer choices
to eliminate.

Example 1 Multiple Choice Tiffany, Lynn, and Rebeckah opened a
lemonade stand to earn extra money. They earned $24.87. How
much money will each girl receive if they split their earnings evenly?

A $6.99   **B** $8.29   **C** $8.33   **D** $9.00

Solution:
Use mental math to estimate the answer:

$24.87 \approx 24$

$\dfrac{24}{3} = 8$ The amount they receive should be slightly more than $8.

Eliminate Choice A since $6.99 is too small.
Eliminate Choice D since $9.00 is too large.
Choose between Choices B and C.

$\dfrac{24.87}{3} = 8.29$

Each girl will receive $8.29.
The correct answer is Choice B.

Example 2 James went clothes shopping at a department store. He
chose 3 shirts, priced at $18.49, $27.89, and $21.39. He also picked
out two pairs of jeans, priced at $28.59 and $32.29. Which is the
closest estimate of the total cost of his purchases, before tax?

F $120   **G** $130   **H** $135   **I** $140

Solution:
Use mental math to estimate the answer:
Add the dollar amounts:
$18 + $27 + $21 + $28 + $32 = $126
Eliminate Choice F because it is too small.
Eliminate Choice I because it is too large.

Choose between Choices G and H.
The correct answer is Choice G.

33   **Holt Middle School Math**   **Course 1**

Name _____ Date _____ Class _____

Test Taking Strategy
Chapter 3, *continued*

Exercises Possible answers are given.

Identify two choices that you can eliminate immediately and explain your reasoning. Then solve the problem.

1. When Mrs. Johnson found money in the laundry, she would divide it evenly between her 4 children. This week she found $35.64. How much will each child receive?

   A $6.73   **B** $8.21   **C** $8.91   **D** $9.32

   Choice 1: Choice A is too small. _____

   Choice 2: Choice D is too large. _____

   The correct calculation is $\frac{35.64}{4} = 8.91$. The correct answer is Choice C.

2. Sally went to the movies on Saturday. She paid $6.75 for her ticket. She bought a soda for $2.39 and a bag of popcorn for $3.65. After the movie, she went to dinner and paid a total of $9.24. Which is the closest estimate to how much Sally spent?

   F $9   **G** $22   **H** $25   **I** $27

   Choice 1: Eliminate Choice F since it is unreasonable. _____

   Choice 2: Eliminate Choice I since $27 is too large. _____

   Choice G is the correct answer. _____

3. Jason and Samantha are planning a ski trip. They recorded the snowfall from the local newspaper. The first day it snowed 5.7 inches. The next four days it snowed 6.2 in., 5.4 in., 7.7 in., and 8.2 in. How much snow fell in total?

   A 29 in.   **B** 30 in.   **C** 33 in.   **D** 77 in.

   Choice 1: Eliminate Choice A since 29 in. is too small. _____

   Choice 2: Eliminate Choice D since 77 in. is unreasonable. The total

   snowfall must be less than 35 in. because if you round up the values you

   get, $6 + 7 + 6 + 8 + 9 = 36$. Choice C is the correct answer.

Holt Middle School Math   Course 1

Test Taking Strategy
Chapter 4

Know How the Test Is Scored

Different standardized tests are scored in different ways. Pay attention to the directions the test proctor reads aloud. It helps if you know ahead of time how the test is scored.

• Some multiple choice sections have no penalty for guessing. In this case, be sure to answer every question.

• If the test has a penalty for guessing and you cannot eliminate any answer choices, it is best to leave the question blank.

• Extended response questions are graded using a scoring rubric. Never leave an extended response question blank. Always show each step of your calculations, and explain your thinking process.

Example Extended Response Every 15 days, Jan visits her grandmother. Every 12 days, she visits her uncle. In how many days will Jan visit her grandmother and her uncle on the same day? Show all of your steps and explain your reasoning.

Scoring Rubric

• **4 points:** Student recognizes the need to find the LCM of 12 and 15. Shows calculations to find LCM. Explains reasoning and answers in a complete sentence.

• **3 points:** Student recognizes the need to find the LCM of 12 and 15 but does not show all steps or does not give a clear explanation.

• **2 points:** Student recognizes the need to find LCM but calculates incorrectly or does not complete the solution.

• **1 point:** Student gives a numeric answer but includes no explanation or work.

• **0 points:** Student gives incorrect solution with no explanation or work included.

Solution: A possible 2-point response is shown below. Notice that although the student calculates the LCM as 120 rather than 60, he still receives 2-points for the response.

Find the LCM of 12 and 15.
12: 12, 24, 36, 48, 60, 72, 84, 96, 108, 120
15: 15, 30, 45, 60, 75, 90, 105, 120

The LCM of 12 and 15 is 120

Holt Middle School Math   Course 1

Test Taking Strategy
Chapter 4, continued

Exercises
Use the scoring rubric to answer the following questions.

Scoring Rubric

- **4 points:** Student recognizes the need to find the total number of cars in the lot. Shows calculations, explains reasoning, and answers in a complete sentence.

- **3 points:** Student recognizes the need to find the total number of cars in the lot but does not show all steps or does not give a clear explanation.

- **2 points:** Student recognizes the need to find the total number of cars, but calculates incorrectly or does not complete the solution.

- **1 point:** Student gives a numeric answer but includes no explanation or work.

- **0 points:** Student gives incorrect solution with no explanation or work included.

---

1. A parking lot has space for 1,000 cars but $\frac{2}{5}$ of the spaces are for compact cars. On Tuesday, there were 250 compact cars and some standard size cars in the parking lot. The parking lot was $\frac{3}{4}$ full. How many standard size cars were in the parking lot?

Solution: 2-point response

   I know that the parking lot is $\frac{3}{4}$ full. $\frac{3}{4} \times 1,000 = 750$.

a. Explain why the student only received 2-points for this response.

   Possible answer: The student failed to show the calculations to arrive

   at the answer of 500 cars.

b. Write a 4-point response.

   I know that the parking lot is $\frac{3}{4}$ full. $\frac{3}{4} \times 1,000 = 750$.

   There are 250 compact cars. $750 - 250 = 500$

   There are 500 standard size cars in the parking lot.

Name _____ Date _____ Class _____

Test Taking Strategy
Chapter 5

Gridded Response

Gridded response questions require that you fill in your answer on the grid provided on the answer sheet.

Response Grids have these parts:

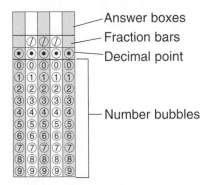

— Answer boxes
— Fraction bars
— Decimal point

— Number bubbles

Follow these steps to grid a decimal or fractional answer:

1. Write your answer in the answer boxes at the top of the grid. Put the first digit of your answer in the box on the left OR put the last digit of your answer in the box on the right.

2. Put only one digit, or the fraction bar, or the decimal point in each box. Do NOT leave a blank box in the middle of an answer. Mixed numbers *cannot* be written in the answer box. You must convert the answer to an improper fraction. So, $2\frac{1}{6}$ must be gridded as 13/6.

3. Remember that the fraction bar and the decimal point have a designated box.

4. Shade the bubble of each digit in the same column as the digit in the answer box.

5. Always use a pencil. Be sure to fill in the entire bubble. Be careful not to rip the paper.

   Example: Grid the answer 0.41.

   Write the number starting with 0 in the first box on the left.

   Shade the decimal point between 0 and 4.

   Shade the correct number bubbles.

Holt Middle School Math   Course 1

Test Taking Strategy
Chapter 5, continued

Exercises

What should go in the second box on the left for each gridded response answer?

1. $\frac{33}{35}$ _____ 3

2. $\frac{16}{25}$ _____ 6

3. $\frac{99}{10}$ _____ 9

4. $34\frac{1}{2}$ _____ 9

Which column should the fraction bar go in for each gridded response answer?

5. $\frac{24}{25}$ _____ third

6. $\frac{1}{500}$ _____ second

7. $\frac{12}{41}$ _____ third

8. $\frac{910}{2}$ _____ fourth

Determine if each value is gridded correctly. Explain.

9. $\frac{9}{10}$

10. $2.77\overline{7}$

11. $15\frac{1}{3}$

12. $1\frac{4}{5}$

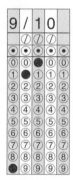

No, the fraction bar was not shaded.

Yes. $2.77\overline{7}$ rounds to 2.778.

No, you cannot shade mixed numbers, need to change to improper fraction first.

No, the 8 was shaded instead of 9.

Grid each value.

13. $\frac{3}{5}$

14. $\frac{17}{18}$

15. $2\frac{1}{3}$

16. $12\frac{1}{5}$

Name _____ Date _____ Class _____

Test Taking Strategy
Chapter 6

Context-Based
Response Questions

Use the information provided in a table, a diagram, or a graph to answer a context-based response question. To receive full credit for this type of question, you need to explain your reasoning and show all of your work in detail. Be sure to check your answer for sense once you complete the question.

Example Wesley displays the following graph in his science class. He then makes the statement that there are more known beetle species than all of the other types of species combined. Do you support Wesley's statement? Why or why not?

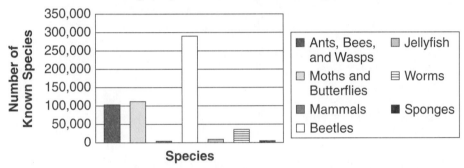

To receive full credit for this question, work through this type of process.

First read over the question again: What information does the question statement provide? Plan to use this information to solve the problem.

Given: Data from the graph can be used to find the number of known species of beetles as well as the total number of known species of all of the other types of species. Then you can compare the two values to determine if you support Wesley's statement.

Find the number of species: Beetles: There are about 290,000 types of beetle species. All other species: 103,000 + 112,000 + 4,000 + 9,000 + 36,000 + 5,000 = 269,000
There are about 269,000 types of other species shown.

Answer in a complete sentence: I support Wesley's statement because 290,000 is greater than 269,000.

Check that your answer is reasonable: By looking at the graph, you can see that the number of known beetle species is far greater than any of the other types of species. It is a reasonable conclusion that Wesley's statement is correct.

Holt Middle School Math   Course 1

Test Taking Strategy
Chapter 6, continued

Exercises

Answer each question.

1. Data for 6 out of the 8 planets is shown below. Maria says the mean length of day in Earth hours is greater than that of the median. Is she correct? Explain.

Planetary Days

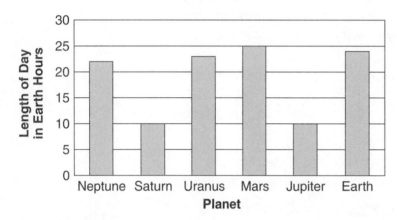

a. What information does the problem statement provide? Write a plan for how you will use this information to answer the question.

a bar graph; I will use the values on the graph to calculate the mean and median of the data. I will then compare the mean and median to determine if Maria is correct.

b. Use your plan of solve the problem. Show your work.

mean = 22 + 10 + 23 + 25 + 10 + 24 = 114 ÷ 6 = 19 Earth hours, median = 22.5 Earth hours; No Maria is not correct. The mean is not greater than the mode.

c. Is your answer reasonable? Explain.

Possible answer: Yes, by looking at the graph there are more larger values than smaller values, so the median would be a larger number. The mean would not be as large as the median because of the two smaller values.

Holt Middle School Math   Course 1

Test Taking Strategy   Short Response—Drawing Diagrams
Chapter 7

Some short response questions require you to draw and label a diagram. You need to make sure you draw the figure as described in the problem statement and provide all markings and labeling as appropriate. Short response questions often have multiple correct solutions. So, your score is based on a scoring rubric.

Example 1
Short Response Sketch and label $\angle ABC$ and $\angle CBD$, a pair of adjacent, supplementary angles.

Scoring Rubric

- **2 points:** Student correctly sketches a pair of adjacent, supplementary angles, and correctly labels them *ABC* and *CBD*.

- **1 points:** Student correctly sketches the angles, but labels them incorrectly, or sketches adjacent angles that are not supplementary.

- **1 point:** Student sketches two angles that are neither adjacent nor supplementary.

- **0 points:** Student sketches one angle or has no response.

Solution
2-point response:
See the sketch at the right for one possible response. Notice that the angles are adjacent to one another. The angles are supplementary, meaning they form a 180° angle, and points *A*, *B*, *C*, and *D* are labeled.

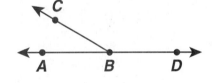

1-point response:
See the sketch at the right for one possible response. Notice that the angles are adjacent to one another, however, they are NOT supplementary, meaning they do NOT form a 180° angle. Although the angles are labeled, they are not labeled correctly. The angles in the diagram are $\angle ACB$ and $\angle BCD$, not $\angle ABC$ and $\angle CBD$.

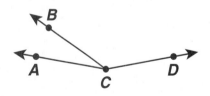

0-point response:
See the sketch at the right for one possible response. Notice that only one angle is drawn and labeled. The student did not complete the required diagram.

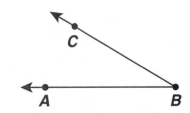

Holt Middle School Math Course 1

Test Taking Strategy
Chapter 7, continued

Exercises Possible answers are given.
Use the scoring rubric for each question.

Scoring Rubric

- **2 points:** Student sketches diagram as required, shows all work and makes no errors in computations. The question is answered in a complete sentence.

- **1 point:** Student shows most work and makes no errors. The student answers the question but fails to draw a diagram.

- **0 points:** There is no response or it is completely incorrect.

---

1. Draw a regular 8-sided polygon Find the sum of the angles measures in the figure and find the measure of each angle.

1-point response	0-point response
$180(8 - 2) = 1,080$ There are 1,080 degrees in a regular 8-sided polygon. $\frac{1,080}{8} = 135$ degrees The measure of each angle is 135°.	$90(8) = 720$ degrees $\frac{720}{10} = 72$ degrees

a. Name three things you would have to add to the 1-point response to make it a 2-point response.

Add a diagram of an 8-sided regular polygon; write the formula for the

sum of the measures of the angles in a polygon, and show all the work.

b. Why did the 0-point response only receive 0-points?

Possible answer: The response is completely incorrect.

2. Sketch and label two right triangles that are congruent. Write two congruence statements. Write a 2-point response.

$\triangle ABC \approx DEF;\ \angle A \approx \angle D;\ \angle C \approx \angle E$

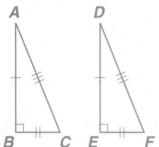

Holt Middle School Math Course 1

Test Taking Strategy
Chapter 8

Response Questions

Short response questions require you to find the solution to a problem but do not provide answer choices or a grid. To get full credit, you need to show each step of your calculations, provide your reasoning, and when appropriate, answer in a complete sentence.

Example 1 Short Response George painted walls at his aunt's house over the weekend. She paid him $39. He worked for 6 hours. How much did George earn per hour?

Solution
Reasoning: Find the unit rate.

Show all of your work: $\frac{39}{6} = 6.5$

Answer the question in a complete sentence: George earned $6.50 per hour.

Example 2 Short Response For a birthday party, Ellen and Abby want to enlarge a 4 in. by 6 in. picture of the birthday girl so that the length of the photo is 15 in. What is the width of the enlargement?

Solution
Reasoning: Set up and solve a proportion.
Show all of your work:

$$\frac{4}{6} = \frac{x}{15}$$
$$4 \cdot 15 = 6 \cdot x$$
$$60 = 6x$$
$$\frac{60}{6} = \frac{6x}{6}$$
$$10 = x$$

Answer the question in a complete sentence:
The width of the enlargement is 10 in.

Test Taking Strategy
Chapter 8, continued

Exercises   Possible answers are given.

Answer each question.

1. Terrance is standing on a sidewalk next to a lamppost. The
   lamppost is 9 ft tall and casts a 3-ft shadow. Terrance is 6 ft tall.
   How long is his shadow?

Response:

$$\frac{9 \text{ ft}}{3 \text{ ft}} = \frac{6 \text{ ft}}{x \text{ ft}}$$

$$9 \cdot x = 3 \cdot 6$$

$$9x = 18$$

$$\frac{9x}{9} = \frac{18x}{9}$$

$$2 = x$$

Terrance's shadow is 2 ft long.

a. The response to the question did not receive full credit. Why?

   The response did not include the student's reasoning.

b. Complete the response so that it receives full credit.

   Write a proportion and solve for *x*, the length of Terrance's shadow.

2. Marcus takes his sister out to dinner for landing the lead role in
   the school play. The cost of the meal is $28. Since the service
   was good, he wants to leave a 20% tip. What is his total bill
   after he adds the tip?

Response:

Calculate the amount of the tip and then add
the amount of the tip to the cost of the meal.

The total bill after the tip is added is $33.60.

The response to the question provides a correct answer. Do you
think the student received full credit for the response? Explain.

No, the student did not show any of the work.

Name _____ Date _____ Class _____

Test Taking Strategy
Chapter 9

Multiple Choice Questions—
Working Backwards

There will be times that you may not know how to solve a multiple choice test question. If the test does not penalize you for guessing, then you need to provide an answer to every question. One method to help you to make an educated guess is to use the answer choices provided and work backwards to solve the question.

Example 1 Find the value of x that makes the equation true.

$x + 4 = -9$

A $x = 4$ **B** $x = -4$ **C** $x = -13$ **D** $x = 13$

Work backwards to find the correct solution.

Try Choice A: $x = 4$; $x + 4 = -9$

$4 + 4 \overset{?}{=} -9$ Substitute 4 into the equation.

$8 \neq -9$ $x = 4$ is not the correct solution.

Try Choice B: $x = -4$; $x + 4 = -9$

$-4 + 4 \overset{?}{=} -9$ Substitute -4 into the equation.

$0 \neq -9$ $x = -4$ is not the correct solution.

Try Choice C: $x = -13$; $x + 4 = -9$

$-13 + 4 \overset{?}{=} -9$ Substitute -13 into the equation.

$-9 = -9$ $x = -13$ is the correct solution.

The correct answer is Choice C.

Example 2 Solve for x. $\frac{x}{3} = -12$

F 36   **G** 4   **H** -4 **I** -36

Try Choice F: $x = 36$

$\frac{36}{3} \overset{?}{=} -12$ Substitute 36 into the equation.

$12 \neq -12$ $x = 36$ is not a correct solution.

Try Choice I next since the only thing wrong with Choice F was the negative sign.

$\frac{-36}{3} \overset{?}{=} -12$ Substitute -36 into the equation.

$-12 = -12$ $x = -36$ is the correct solution.

The correct answer is Choice I.

Holt Middle School Math   Course 1

Test Taking Strategy
Chapter 9, continued

Exercises   Possible answers are given.

1. Find the value of x that makes the equation true. $x + 5 = -4$

A $x = 9$ **B** $x = 1$ **C** $x = -1$ **D** $x = -9$

A student worked backwards to answer the question. The following is the student's work.

Try Choice A: Substitute 9 for x.

$(9) + 5 \stackrel{?}{=} -4;$
$14 \neq -4$

Try Choice C: Substitute -1 for x.

$(-1) + 5 \stackrel{?}{=} -4;$
$4 \neq -4$

Try Choice B: Substitute 1 for x.

$(1) + 5 \stackrel{?}{=} -4;$
$6 \neq -4$

Try Choice D: Substitute -9 for x.

$(-9) + 5 \stackrel{?}{=} -4;$
$-4 = -4$

a. After the student tried Choice A, why did she try another answer choice?

The value in Choice A did not make the equation a true statement.

b. The student selected Choice C as the correct answer. Do you agree with her selection? If not, what selection would you have made? Explain.

No, the student should have chosen Choice D. The only value that

makes the equation a true statement is given in Choice D.

2. Find the value of x that makes the equation true.

$6x + 3 = -21$

F $x = 6$ **G** $x = 3$ **H** $x = -4$ **I** $x = -6$

a. Explain how you can work backwards to answer this question.

Substitute the given value in each answer choice into the equation

and see if it results in a true statement.

b. Once you find one answer choice that is correct, do you need to check the other answer choices? Why?

No, because there is only one correct answer to this question.

Holt Middle School Math   Course 1

Test Taking Strategy
Chapter 10

Sketch a Picture or Diagram

Sketching a picture or diagram can help you find the solution to some problems.

Example 1 Short Response Daniel is planning to fence in a circular garden. The distance from the center of the garden to the edge of the garden is 10 ft. How many feet of fencing should Daniel purchase?

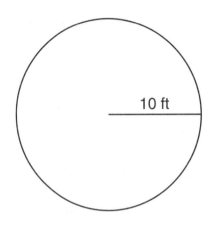

Solution: It helps to draw a diagram with the given information. This way you can visualize the problem.

Sketch a circle with radius labeled "10 ft".

Now calculate the circumference of the circle. Use $\pi \approx 3.14$.

$$C = 2\pi r$$
$$= 2 \cdot \pi \cdot 10$$
$$\approx 62.8$$

Daniel needs to buy 62.8 ft of fencing.

Example 2 Multiple Choice Joseph wants to make a poster shaped like a house. He has a piece of cardboard shaped like a rectangle. The length of the cardboard is 24 in. and its width is 15 in. He also has a piece of cardboard shaped like an isosceles triangle. The base of this cardboard measures 24 in. and the lengths of the other two sides measure 17 in. each. What is the perimeter of the figure after he connects the cardboard together?

A 71 in.   **B** 88 in.   **C** 112 in.   **D** 136 in.

Solution: Draw a diagram to help you understand the situation. Sketch a rectangle with side lengths 15 and 24. Attach to it an isosceles triangle with side lengths 17, 17, and 24.

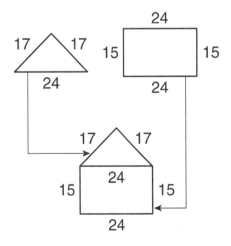

The perimeter of Joseph's poster is calculated as follows:
17 + 17 + 15 + 24 + 15 = 88.

The perimeter of Joseph's poster is 88 in. The correct answer is Choice B.

Holt Middle School Math   **Course 1**

Test Taking Strategy
Chapter 10, continued

Exercises

1. A circle with a radius of 3 inches is inscribed in a square. What is the area between the circle and the square?

   A 36 in^2 **B** 19.26 in^2 **C** 7.74 in^2 **D** 6 in^2

   a. How would you draw a circle *inscribed* in a square?

   Draw a circle inside of a square. _____

   b. Sketch a diagram of the problem.

   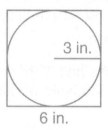

   c. How does the diagram help you to solve the problem?

   Once I drew the diagram I could see the area that needed to be

   calculated.

2. Maria is wrapping a birthday present. The shipping box measures 10 cm by 5 cm by 5 cm. Inside the shipping box is a small gift box that measures 2 cm by 2 cm by 2 cm. How much space is left inside the shipping box for packaging material?

   F 250 cm^2 **G** 242 cm^2 **H** 8 cm^2 **I** 6 cm^2

   a. What geometric shape is the shipping box?

   a rectangular prism _____

   b. Sketch and label a diagram of the problem.

   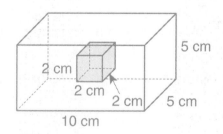

   c. Explain how to determine the correct answer. What is the correct answer?

   I subtracted the volume of the cube from the volume of the box.

   The correct answer is Choice G.

Test Taking Strategy
Chapter 11

Gridded Response

Some questions on standardized tests require you to grid in your answer on a grid. Be sure you know how to correctly grid your answer.

Example Gridded Response There are 33 students in art class. The janitor walks into the class and chooses one student to help him clean the kiln. Eleven of the students are involved in creating a self-portrait. The rest of the class is working on a sculpture. What is the probability of the janitor selecting a student working on the sculpture?

Solution:

There are 33 students in the class and 11 of them are working on a self-portrait. Find the number of students working on the sculpture. $33 - 11 = 22$ students

Next, find the probability of selecting a student working on the sculpture.

P(selecting a student working on the sculpture) $= \frac{22}{33} = \frac{2}{3}$

• You now need to grid in the answer $\frac{2}{3}$.

• When your answer is in the form of a fraction, such as $\frac{2}{3}$, in order to grid your answer correctly, you need to either grid the fraction $\frac{2}{3}$, or convert the fraction to its decimal equivalent, $\frac{2}{3} = 0.\overline{6} \approx 0.666 \approx 0.667$. If a fraction converts to a repeating decimal, every space in the grid must be filled in. You can correctly bubble either 0.666 or 0.667

• Four ways to correctly grid in the answer $\frac{2}{3}$ are shown below.

Test Taking Strategy
Chapter 11, continued

Exercises

What should go in the first box on the left for each gridded response answer?

1. 2.67 __2__   **2.** $\frac{7}{10}$ __7__   **3.** 519 __5__   **4.** $4\frac{1}{2}$ __9__

If you grid your answer starting in the far left column, which column should the decimal or fraction bar go in for each answer below?

5. $\frac{2}{7}$ _second_   **6.** 15.24 _third_   **7.** $3\frac{2}{8}$ _third_   **8.** 895 _none_

Tell what error was made in each gridded response below.

9.

The fourth and fifth columns have zero shaded, they should be blank.

10.

Cannot shade mixed numbers, need to change to improper fraction first.

11.

6 and the fraction bar were shaded in the same column.

12.

0.55

The fourth column should have a 5 shaded in.

Grid each answer.

13. Elizabeth saved money to purchase a new television. She can also choose two of the following options: VCR, DVD, or CD player. How many options does Elizabeth have?

Answer: 3

14. Billy rolls a number cube and flips a coin. Find the probability Billy will roll a 5 and the coin will land on "tails".

Answer: $\frac{1}{12}$

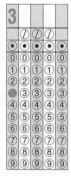

Holt Middle School Math   Course 1

Name _____ Date _____ Class _____

Test Taking Strategy
Chapter 12

Use tables, diagrams, and graphs to answer context-based questions that involve patterns and reasoning.

Example The table shows how much Oscar earns for washing windows. Write an equation representing the amount Oscar earns for washing x windows. Use your equation to determine how much Oscar will earn if he washes 10 windows. Show all of your work.

Number of Windows, x	Amount Earned, y
1	$23
2	$31
3	$39
4	$47
5	$55

First, read over the question again. What information does the question statement provide? Use this information to solve the problem.

Given: A table of values showing the numbers of windows washed and the amount of money earned. You can use the table to determine a function. Then substitute the value of 10 windows into the function to determine the amount of money Oscar will earn.

Determine the function: Look for a pattern: The y-values increase by $8 per window. The function is $y = 8x + 15$, where y is the amount of money earned, and x is the number of windows washed.

Number of Windows, x	Amount Earned, y
1	$8(1) + 15 = 23$
2	$8(2) + 15 = 31$
3	$8(3) + 15 = 39$
4	$8(4) + 15 = 47$
5	$8(5) + 15 = 55$

Use the function:
Substitute 10 for x in the function.
$y = 8x + 15$
$ = 8(10) + 15$
$ = 80 + 15$
$ = 95$ Oscar will earn $95 for washing 10 windows.

Check your answer, does it make sense?
Yes, if you extend the table to 10 windows, and use the pattern, the total amount earned would be $95.

Holt Middle School Math   Course 1

Name _____ Date _____ Class _____

Test Taking Strategy
Chapter 12, continued

Exercises Possible answers are given.

1. Jerry is a bicycle salesman. He gets $50 per day and $35 for every bicycle he sells. If Jerry sells 6 bicycles the first day and 7 bicycles the second day, how much money will he make?

Number of Bicycles, x	Amount Earned, y
1	$85
2	$120
3	$155
4	$190
5	$225

a. Read the problem carefully. What information are you given? What does the question ask?

A table showing the numbers of bicycles sold and the amount of money earned. The amount of money Jerry will make if he sells 6 bicycles the first day and 7 the second day.

b. Write a plan for how you will use the given information to answer the question.

You can use the table to determine a function. Substitute 6 into the function to determine the amount of money Jerry will earn his first day. Substitute 7 into the function to determine the amount of money Jerry will earn his second day. Then add the two amounts together to determine the total money earned.

c. What pattern do you notice in the table? Use the pattern to write a function.

The y-values increase by $35 for every bicycle sold. Let x equal the number of bicycles sold in one day. Let y equal the amount of money earned. $35x + 50 = y$

Holt Middle School Math   Course 1

Answer Key

Chapter 1

1. I can substitute each given answer choice into the pattern and see if that value is part of the sequence.

2. Each value is four less than the previous value. The missing number is 34, Choice C.

3. Choice F and Choice I; The pattern only includes the numbers 4 and 6 and the choices given in F and I cannot be arrived at by a simple mistake or misjudgment.

4. Choice G is a distracter because you could possible make a minor error in calculating the 12 value of the sequence, resulting in the answer of 4.

Chapter 2

1. Possible answer: The student failed to show each step of the work. The student also failed to provide reasoning or write the answer in a complete sentence.

2. C

Chapter 3

1. Choice 1: Possible answer: Choice A is too small.
   Choice 2: Possible answer: Choice D is too large.
   The correct calculation is $\frac{35.64}{4} = 8.91$.
   The correct answer is choice C.

2. Choice 1: Eliminate Choice F since it is unreasonable.
   Choice 2: Eliminate Choice I since $27 is too large.
   Choice G is the correct answer.

3. Choice 1: Eliminate Choice A since 29 in. is too small.
   Choice 2: Eliminate Choice D since 77 in. is unreasonable. The total snowfall must be less than 35 in. because if you round up the values you get, $6 + 7 + 6 + 8 + 9 = 36$. Choice C is the correct answer.

Holt Middle School Math   Course 1

Answer Key

Chapter 4

1. **a.** Possible answer: The student failed to show the calculations to arrive at the answer of 500 cars. The student failed to give a complete sentence.

   b. I know that the parking lot is $\frac{3}{4}$ full. $\frac{3}{4} \times 1{,}000 = 750$. There are 250 compact cars. $750 - 250 = 500$. There are 500 standard size cars in the parking lot.

   c. The student would have only received 1-point.

Chapter 5

1. 3
2. 6
3. 9
4. 9
5. third
6. second
7. third
8. fourth
9. No, the fraction bar was not shaded.
10. Yes. $2.77\overline{7}$ rounds to 2.778.
11. No, you cannot shade mixed numbers, need to change to improper fraction first.
12. No, the 8 was shaded instead of 9.

13.
14.
15.
16.

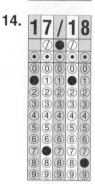

Holt Middle School Math   Course 1

Answer Key

Chapter 6

1. I will use the values on the graph to calculate the mean and median of the data. I will then compare the mean and median to determine if Maria is correct.

   b. mean = 22 + 10 + 23 + 25 + 10 + 24 = 114 ÷ 6 = 19 Earth hours, median = 22.5 Earth hours; No Maria is not correct. The mean is not greater than the median.

   c. Possible answer: Yes, by looking at the graph there are more larger values than smaller values, so the median would be a larger number. The mean would not be as large as the median because of the two smaller values.

Chapter 7

1. Add a diagram of an 8-sided regular polygon; write the formula for the sum of the measures of the angles in a polygon, and show all the work.

   b. Possible answer: The response is completely incorrect.

2.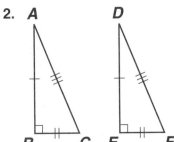

$\triangle ABC \approx DEF$; $\angle A \approx \angle D$; $\angle C \approx \angle E$

Chapter 8

1. The response did not include the student's reasoning.

   b. Write a proportion and solve for *x*, the length of Terrance's shadow.

2. No, the student did not show any of the work.

Chapter 9

1. a. The value in Choice A did not make the equation a true statement.

   b. No, the student should have chosen Choice D. The only value that makes the equation a true statement is given in Choice D.

2. Substitute the given value in each answer choice into the equation and see if it results in a true statement.

   b. No, because there is only one correct answer to this question.

Chapter 10

1. Draw a circle inside of a square.

   b.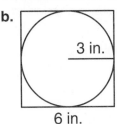

   c. Once I drew the diagram I could see the area that needed to be calculated.

2. a rectangular prism

   b.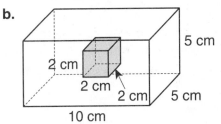

   c. I subtracted the volume of the cube from the volume of the box. The correct answer is Choice G.

Answer Key

Chapter 11

1. 2
2. 7
3. 5
4. 9
5. second
6. third
7. third
8. none
9. The fourth and fifth columns have zero shaded, they should be blank.
10. Cannot shade mixed numbers, need to change to improper fraction first.
11. 6 and the fraction bar were shaded in the same column.
12. The fourth column should have a 5 shaded in.

13.

14.

Chapter 12

1. **a.** A table showing the numbers of bicycles sold and the amount of money earned. The amount of money Jerry will make if he sells 6 bicycles the first day and 7 the second day.

   b. You can use the table to determine a function. Substitute 6 into the function to determine the amount of money Jerry will earn his first day. Substitute 7 into the function to determine the amount of money Jerry will earn his second day. Then add the two amounts together to determine the total money earned.

   c. The y-values increase by $35 for every bicycle sold. Let x equal the number of bicycles sold in one day. Let y equal the amount of money earned. $35x + 50 = y$

Holt Middle School Math   Course 1

Standardized Test Practice

Chapter 1

Select the best answer for Questions 1–6.

1. Which is a true statement?

   A $34,567 > 35,678$

   B $2,345 > 3,254$

   C $123,984 > 122,894$

   D $813,432 < 812,423$

2. Davey's Food Service has provided the food and supplies for a local road race the last few years. Use the table to estimate the number of cups the road race will use in 4 years.

Davey's Food Service	
Item	**Number Supplied per Year**
Sports Drink	75 gallons
Cups	534
Energy Bars	185

   F 500 cups

   G 1,500 cups

   H 2,000 cups

   I 2400 cups

3. Five men each buy 12 plastic bats for $2 each. How much did they spend?

   A $10

   B $22

   C $120

   D $240

4. Everyday Wylie picks up aluminum cans. On day one he picked up two cans. Each day after that he picked up twice as many cans as the day before. Which expression would you use to find the number of cans he picked up on the sixth day?

   F 1^6

   G 2^6

   H 2^4

   I 6^2

5. There are 4 people in the Marker family. Each person takes about an 8-minute shower. The family wants to conserve water due to a drought. They know that their showerhead uses about 3 gallons of water per minute. How many gallons of water does the family use to take showers each day?

   A 24 gallons

   B 48 gallons

   C 96 gallons

   D 112 gallons

6. There are 6 teachers and 244 students traveling to the art museum for a field trip. One bus will hold 40 people. How many buses are needed altogether?

   F 4 buses

   G 5 buses

   H 6 buses

   I 7 buses

Holt Middle School Math   Course 1

Standardized Test Practice

Chapter 1, continued

Gridded Response

Solve the problems. Use the answer sheet to write and grid-in your answer.

7. Nicholas sold a baseball card for $52. Its value had increased $2 each of the six years he owned it. How much did Nicholas originally pay for the card?

8. What is the area, per 1,000 square kilometers, of the continent with the smallest area?

Size of the Continents	
Continents	**Size (sq km x 1,000)**
Africa	30,065
Asia	44,579
Antarctica	13,209
Australia	7,687
Europe	9,938
North America	24,256
South America	17,819

9. Laura's family took a family vacation. The first day they drove 420 miles; which was one third of the distance of the entire trip. The next day the car had a flat tire, so they were only able to drive 60 miles. How many more miles does Laura's family have to drive?

Short Response

Solve the problems. Use the answer sheet to write your answers.

10. There are 15 players on a soccer team. Each player receives a jersey, shorts, and socks. A jersey costs $15, shorts cost $12, and socks cost $3. Write and evaluate an expression for the total cost of the team's outfit.

11. Harriet evaluated the following problem on a quiz and the teacher marked it as incorrect. Explain in words what Harriet did wrong. What is the correct answer?

$$3 + 4 \times 6 + 8 = 7 \times 14 = 98$$

Extended Response

12. In a process called *mitosis*, a cell divides to form two new cells. Then those new cells divide again to form new cells and those new cells divide yet again to form more new cells, etc.

a. Start with one cell and write how many cells there will be after the first division, the second division, and the third division. Make a table to show the pattern.

b. Explain the pattern that you notice.

c. If there are 56 cells during a stage of mitosis, how many cells will there be during the next stage? Explain in words how you determined your answer.

Holt Middle School Math   Course 1

Standardized Test Practice Answer Sheet

Chapter 1

Multiple Choice

1. (A) (B) (C) (D) See Lesson 1-1.

2. (F) (G) (H) (I) See Lesson 1-2.

3. (A) (B) (C) (D) See Lesson 1-5.

4. (F) (G) (H) (I) See Lesson 1-3.

5. (A) (B) (C) (D) See Lesson 1-6.

6. (F) (G) (H) (I) See Lesson 1-2.

Gridded Response

7.

See Lesson 1-7.

8.

See Lesson 1-1.

9.

See Lesson 1-6.

Short Response

Write your answers in the space provided.

10. _____

_____

_____

(See Lesson 1-4.)

11. _____

_____

_____

_____

(See Lesson 1-4.)

Extended Response

Write your answers for Problem 12 on the back of this paper.

See Lesson 1-7.

Holt Middle School Math   Course 1

Name _____ Date _____ Class _____

Standardized Test Practice Answer Sheet
Chapter 1, continued

12. Part A.

Part B. _____

Part C. _____

_____

_____

Holt Middle School Math   Course 1

Standardized Test Practice Answer Sheet

Chapter 1

Multiple Choice

1. Ⓐ Ⓑ ● Ⓓ See Lesson 1-1.

2. Ⓕ Ⓖ ● Ⓘ See Lesson 1-2.

3. Ⓐ Ⓑ ● Ⓓ See Lesson 1-5.

4. Ⓕ ● Ⓗ Ⓘ See Lesson 1-3.

5. Ⓐ Ⓑ ● Ⓓ See Lesson 1-6.

6. Ⓕ Ⓖ Ⓗ ● See Lesson 1-2.

Gridded Response

7.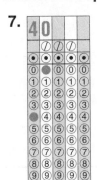

See Lesson 1-7.

8.

See Lesson 1-1.

9.

See Lesson 1-6.

Short Response

Write your answers in the space provided.

10. $15(15 + 12 + 3) = 15(30) = 450$; The total cost of the team's uniforms is $450.

(See Lesson 1-4.)

11. Harriet added first. She should have multiplied the 4 and 6 first and then added.

$3 + 4 \times 6 + 8 = 3 + 24 + 8 = 35$

The correct answer is 35.

(See Lesson 1-4.)

Extended Response

Write your answers for Problem 12 on the back of this paper.

See Lesson 1-7.

Holt Middle School Math   Course 1

Standardized Test Practice Answer Sheet
Chapter 1, continued

12. Part A.

Number of Divisions	Number of Cells
0	1
1	2
2	4
3	8

Part B. After each division the number of cells doubles.

Part C. If there are 56 cells and the number of cells doubles after each mitosis, then there will be 56 × 2 cells or 112 cells in the next stage.

Scoring Rubric

4 The student correctly identifies the number of cells in the first three stages, describes the pattern, predicts the number of cells in a following stage, and provides an explanation of the solution.

3 The student correctly identifies the number of cells in the first three stages, correctly describes the pattern, but incorrectly predicts the number of cells in a following stage.

2 The student correctly identifies the number of cells in the first three stages, but does not correctly describe the pattern, thus incorrectly predicting the number of cells in a following stage.

1 The student correctly identifies the pattern, but does not describe the pattern and does not attempt to predict the number of cells in a following stage.

0 The answers are not correct and no work is shown.

Standardized Test Practice

Chapter 2

Select the best answer for Questions 1–6.

1. Mario's truck holds 26 gallons of gas. The tank had 8 gallons in it, and Mario added 18 gallons when he fueled at the gas station. Which equation best represents this situation?

   A $8 + 18 = 26$

   B $18 - 8 = 26$

   C $18(8) = 26$

   D $\frac{26}{8} = 18$

2. Which expression is represented in the table?

n	?
18	3
24	4
30	5

   F $n - 15$

   G $6n$

   H $\frac{n}{6}$

   I $n + 15$

3. A farmer picks 87 pumpkins and puts them in a wagon to sell at the side of a road. If at the end of the day, he has 13 pumpkins left, how many pumpkins did he sell?

   A 64 pumpkins

   B 74 pumpkins

   C 78 pumpkins

   D 100 pumpkins

4. A football play starts on the 18 yard line and ends on the 42 yard line. How many yards were gained in the play?

   F 24 yards

   G 60 yards

   H 34 yards

   I 756 yards

5. There are usually 4 tiger cubs in a litter. If a tiger has 16 cubs in her lifetime, how many litters did she probably have?

   A 2 litters

   B 4 litters

   C 12 litters

   D 64 litters

6. During the month of November, Kendra saved d dollars. During the month of December, she saved $4.00 more than the amount that she saved in November. Which expression represents the amount of money Kendra saved in December?

   F $d + \$4.00$

   G $d - \$4.00$

   H $\$4.00 \times d$

   I $\frac{\$4.00}{d}(25)$

Holt Middle School Math   Course 1

Name _____ Date _____ Class _____

Standardized Test Practice
Chapter 2, continued

Gridded Response
Solve the problems. Use the answer sheet to write and grid-in your answer.

7. During a class activity you and your partner measure your heights in inches. Your total combined height is 114 inches. If you are 56 inches tall, how many inches tall is your partner?

8. An entire cake has 340 calories. If you cut the cake into 10 pieces, how many calories does each piece have?

9. You have $50 in the bank and each week you add $20 to your account. The expression $50 + 20x$ represents the amount of money you will save after x weeks. How many dollars will you have in 8 weeks?

10. Maria baked cookies for her daughter's class. There are 18 students in her daughter's class plus a teacher and an educational aide. If each person gets 3 cookies, how many cookies did Maria bake?

Short Response
Solve the problems. Use the answer sheet to write your answers.

11. Your math partner says that in order to solve $5x = 105$ you subtract 5 from each side. Do you agree? Explain your reasoning.

12. Write an expression for the phrase twice the sum of a number, x, and fifteen. Explain in words how to evaluate the expression if $x = -6$, then evaluate the expression.

13. Christian wants to buy some oranges for 50 cents each. He has $2. How many oranges can he buy? Write and solve an equation.

Extended Response

14. Use the figure.

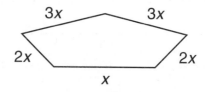

a. Write an expression for the perimeter of the pentagon.

b. If $x = 3$ in., what is the measure of each side? Show two different ways to calculate the perimeter.

c. If the perimeter is 132 inches, what is x? What is the measure of each side? Show your work.

Holt Middle School Math   Course 1

Standardized Test Practice Answer Sheet
Chapter 2

Multiple Choice

1. (A) (B) (C) (D) See Lesson 2-3.
2. (F) (G) (H) (I) See Lesson 2-1.
3. (A) (B) (C) (D) See Lesson 2-5.

4. (F) (G) (H) (I) See Lesson 2-3.
5. (A) (B) (C) (D) See Lesson 2-7.
6. (F) (G) (H) (I) See Lesson 2-2.

Gridded Response

7.

See Lesson 2-4.

8.

See Lesson 2-6.

9.

See Lesson 2-1.

10.

See Lesson 2-7.

Short Response
Write your answers in the space provided.

11. _____
_____
_____ (See Lesson 2-6.)

12. _____
_____
_____ (See Lesson 2-2.)

13. _____
_____
_____ (See Lesson 2-12.)

Extended Response
Write your answers for Problem 14 on the back of this paper.
See Lesson 2-1.

Name _____ Date _____ Class _____

Standardized Test Practice Answer Sheet
Chapter 2, continued

14. Part A. _____

Part B. _____

_____

Part C. _____

_____

Holt Middle School Math   Course 1

Name _____ Date _____ Class _____

Standardized Test Practice Answer Sheet

Chapter 2

Multiple Choice

1. ● Ⓑ Ⓒ Ⓓ See Lesson 2-3.
2. Ⓕ Ⓖ ● Ⓘ See Lesson 2-1.
3. Ⓐ ● Ⓒ Ⓓ See Lesson 2-5.

4. ● Ⓖ Ⓗ Ⓘ See Lesson 2-3.
5. Ⓐ ● Ⓒ Ⓓ See Lesson 2-7.
6. ● Ⓖ Ⓗ Ⓘ See Lesson 2-2.

Gridded Response

7.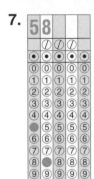

See Lesson 2-4.

8.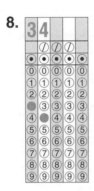

See Lesson 2-6.

9.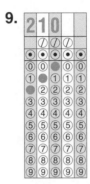

See Lesson 2-1.

10.

See Lesson 2-7.

Short Response

Write your answers in the space provided.

11. No; The equation says 5 times a number is 105. To find the number you have to divide both sides of the equation by 5.

(See Lesson 2-6.)

12. $2(x + 15)$; Substitute the value -6 into the expression for x. Then use the order of operations to evaluate the expression. $2(-6 + 15) = 18$

(See Lesson 2-2.)

13. $0.5x = 2$; $x = 4$; Christian can buy 4 oranges.

(See Lesson 2-12.)

Extended Response

Write your answers for Problem 14 on the back of this paper.
See Lesson 2-1.

Holt Middle School Math   Course 1

Standardized Test Practice Answer Sheet
Chapter 2, continued

14. Part A. $3x + 3x + 2x + 2x + x = 11x$

Part B. $3(3) = 9$ in.; $3(3) = 9$ in.; $2(3) = 6$ in.; $2(3) = 6$ in.; 3 in.;

$11(3) = 33$ in. or $3(3 + 3 + 2 + 2 + 1) = 33$ in.

Part C. $11x = 132$; $x = 12$ in.; 36 in., 36 in., 24 in., 24 in., 12 in

Scoring Rubric

4 The student shows all the work and has complete and correct answers for all three parts.

3 The student shows all the work and has correct answers, but only shows one way to correctly calculate the perimeter of the figure.

2 The student writes an expression to correctly determine the perimeter, but does not give any explanation on how to calculate the perimeter figure.

1 The student shows some work to correctly determine the perimeter, but does not identify the expression and incorrectly shows how to calculate the perimeter.

0 The answers are not correct and no work is shown.

Name _____ Date _____ Class _____

Standardized Test Practice
Chapter 3

Select the best answer for Questions 1–7.

1. Martins' Media Shop is selling three DVD's for $63.75. What is the price of each DVD?

   A $63.75

   B $60.75

   C $21.25

   D $19.12

2. You are stacking 110-pound boxes on a freight elevator. A sign on the elevator says, "Do not exceed 1,200 pounds." What is the maximum number of boxes you can stack on the elevator?

   F 8 boxes

   G 9 boxes

   H 10 boxes

   I 11 boxes

3. Mr. Nye wants to purchase 25 pairs of headphones for the computer lab. If the headphones are on sale for $4.99, about how much will it cost to buy the headphones?

   A $30   **C** $125

   B $50   **D** $200

4. Sidney spent $13.12 on bottled water. If bottled water cost $0.82 per gallon, how many gallons of water did Sidney buy?

   F 4 gallons

   G 10 gallons

   H 12 gallons

   I 16 gallons

5. The school relay team competed at the district meet. The runners times are shown below. What was the total time (in seconds) for the relay team?

Relay Team Results	
Runner	Time (seconds)
Marcus	20.2
Jose	19.3
Roberto	18.7
Steven	16.2

   A 18.6 seconds

   B 20.2 seconds

   C 37.2 seconds

   D 74.4 seconds

6. Stella works at a veterinarian's office 15 hours per week. If she earns $7.25 per hour, how much will she earn in one week?

   F $2.10

   G $7.75

   H $22.25

   I $108.75

7. What is the length of a, in meters?

   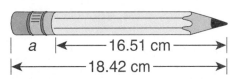

   |← a →|← 16.51 cm →|
   |← 18.42 cm →|

   A 0.3493 m

   B 0.304 m

   C 0.0211 m

   D 0.0191 m

Holt Middle School Math   Course 1

Standardized Test Practice

Chapter 3, continued

Gridded Response

Solve the problems. Use the answer sheet to write and grid-in your answer.

8. When estimating the distance from the earth to the sun, Jenny needs to write the number 149,600,000 km in scientific notation. What exponent should Jenny use on base 10?

9. Mary went to lunch with $8.50. She bought an apple for $0.65, a bottle of water for $0.75, a turkey sandwich for $3.75, and a yogurt for $0.80. How much money should Mary have left?

10. As a lab team, you and your three partners measured the same piece of string. The measurements were 3.011 in., 3.01 in., 3.02 in., and 3.012 in. Which measurement was the smallest?

Short Response

Solve the problems. Use the answer sheet to write your answers.

11. Jill wrote a check for $21.18 and then had $115.62 left in her checking account. Write an equation to determine how much she had in her account before she wrote the check, then solve the equation.

12. Mrs. Henderson's math class ordered pizza for lunch so that they could stay in and study for their unit exam. They ordered 8 pizzas and six 2-liters of soda. The total cost, including tip, was $76.25. If 25 people chipped in the same amount of money, how much did each person pay? Show your work.

Extended Response

13. The longest side of the triangle is 21 units long.

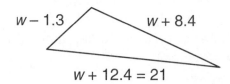

$w - 1.3$ $w + 8.4$

$w + 12.4 = 21$

a. Explain in words how to determine the lengths of the other two sides of the triangle.

b. Find the unknown lengths.

c. What is the perimeter of the triangle? Show your work.

Holt Middle School Math   Course 1

Standardized Test Practice Answer Sheet
Chapter 3

Multiple Choice

1. (A) (B) (C) (D)   See Lesson 3-7.   5. (A) (B) (C) (D)   See Lesson 3-3.

2. (F) (G) (H) (I)   See Lesson 3-9.   6. (F) (G) (H) (I)   See Lesson 3-6.

3. (A) (B) (C) (D)   See Lesson 3-2.   7. (A) (B) (C) (D)   See Lesson 3-4.

4. (F) (G) (H) (I)   See Lesson 3-8.

Gridded Response

8.

See Lesson 3-5.

9.

See Lesson 3-3.

10.

See Lesson 3-1.

Short Response

Write your answers in the space provided.

11. _____

_____

_____ (See Lesson 3-10.)

12. _____

_____

_____

_____ (See Lesson 3-7.)

Extended Response

Write your answers for Problem 13 on the back of this paper.

See Lesson 3-10.

Holt Middle School Math   Course 1

Name _____ Date _____ Class _____

Standardized Test Practice Answer Sheet

Chapter 3, continued

13. Part A. _____

_____

_____

Part B. _____

_____

Part C. _____

Holt Middle School Math   Course 1

Name _____ Date _____ Class _____

Standardized Test Practice Answer Sheet

Chapter 3

Multiple Choice

1. Ⓐ Ⓑ ⬤ Ⓓ   See Lesson 3-7.
2. Ⓕ Ⓖ ⬤ Ⓘ   See Lesson 3-9.
3. Ⓐ Ⓑ ⬤ Ⓓ   See Lesson 3-2.
4. Ⓕ Ⓖ Ⓗ ⬤   See Lesson 3-8.

5. Ⓐ Ⓑ Ⓒ ⬤   See Lesson 3-3.
6. Ⓕ Ⓖ Ⓗ ⬤   See Lesson 3-6.
7. Ⓐ Ⓑ Ⓒ ⬤   See Lesson 3-4.

Gridded Response

8.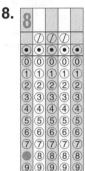

See Lesson 3-5.

9.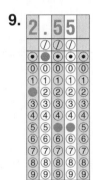

See Lesson 3-3.

10.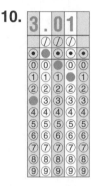

See Lesson 3-1.

Short Response

Write your answers in the space provided.

11. $x - 21.18 = 115.62$; $x = 115.62 + 21.18$; $x = 136.80$; She had $136.80 in her account.

(See Lesson 3-10.)

12.
```
      3.05
 25)76.25
    75
    125
    125
```
; Each person chips in $3.05.

(See Lesson 3-7.)

Extended Response

Write your answers for Problem 13 on the back of this paper.
See Lesson 3-10.

Holt Middle School Math   Course 1

Standardized Test Practice Answer Sheet

Chapter 3, continued

---

13. Part A. Solve for *w* in the equation $w + 12.4 = 21$. Then substitute the value for *w* into each expression to determine the lengths of the triangles sides.

Part B. $w + 12.4 = 21$; $w = 8.6$; $8.6 - 1.3 = 7.3$; $8.6 + 8.4 = 17$; The lengths of the unknown sides are 7.3 units and 17 units.

Part C. $21 + 7.3 + 17 = 45.3$; The perimeter is 45.3 units.

Scoring Rubric

4   The student calculates the other two side lengths and the perimeter correctly and shows and explains all the work.

3   The student shows all the work and the explanation is correct, but has minor calculation errors in finding the perimeter.

2   The student shows some of the work, but has minor calculation errors in finding the perimeter and the explanation is not complete.

1   The student shows some of the work and has calculation errors throughout the exercise. There is no explanation of the work shown.

0   The answers are not correct and no work is shown.

Holt Middle School Math   Course 1

Standardized Test Practice

Chapter 4

Select the best answer for Questions 1–7.

1. For Nina's birthday party, her parents buy a piñata. If there will be a total of nine children at the party, which size package of party favors should her parents buy so that all the children will get an equal amount?

    A 48 favors

    B 97 favors

    C 116 favors

    D 135 favors

2. After two minutes of biking, Brad had biked $\frac{1}{2}$ kilometer, Harold $\frac{3}{4}$ kilometer, Kylie $\frac{4}{5}$ kilometer and Lyle $\frac{2}{3}$ kilometer. Who had biked the farthest?

    F Brad

    G Harold

    H Kylie

    I Lyle

3. Order the fractions and decimals from least to greatest.

    $0.3, \frac{2}{7}, \frac{1}{3}, 0.37$

    A $\frac{2}{7}, 0.3, \frac{1}{3}, 0.37$

    B $0.3, \frac{2}{7}, 0.37, \frac{1}{3}$

    C $\frac{1}{3}, \frac{2}{7}, 0.3, 0.37$

    D $\frac{2}{7}, \frac{1}{3}, 0.3, 0.37$

4. The 6th grade basketball team, have scored 41, 23, 39, and 47 points at their last four games. Which score is a composite number?

    F 23

    G 39

    H 41

    I 47

5. A recipe for your favorite dessert calls for $\frac{2}{3}$ cup of cocoa. How much do you need if you want to make a double batch?

    A $\frac{3}{4}$ cup

    B $1\frac{1}{4}$ cup

    C $1\frac{1}{3}$ cup

    D $1\frac{2}{3}$ cup

6. There are 12 players on the Braves youth baseball team and only $\frac{3}{4}$ of them showed up for practice on Monday night. How many players came to practice?

    F 3 players   **H** 9 players

    G 6 players   **I** 11 players

7. In geometry the fraction $\frac{22}{7}$ is often used as an approximation for pi. What is this improper fraction as a mixed number?

    A $\frac{7}{22}$ **C** $2\frac{2}{7}$

    B $1\frac{2}{7}$ **D** $3\frac{1}{7}$

Holt Middle School Math   **Course 1**

Standardized Test Practice

Chapter 4, continued

Gridded Response

Solve the problems. Use the answer sheet to write and grid-in your answer.

8. What missing number would make the fractions equivalent?

   $\frac{32}{48} = \frac{8}{?}$

9. Nicole is having a wedding shower for her best friend. She is making balloon centerpieces for each table. She has 48 green balloons and 64 white balloons. If she would like the centerpieces to have the greatest equal number of each color balloon, how many centerpieces can she make?

10. What number has the prime factorization of $2^3 \times 3^2 \times 7 \times 13$?

Short Response

Solve the problems. Use the answer sheet to write your answers.

11. Eleanor is making candle gift baskets. She has 12 green votives, 16 blue votives, and 24 yellow votives. If she would like the baskets to have the greatest equal number of each color in them, how many baskets will Eleanor be able to make? Explain in words how you determined your answer.

12. In the picture, 6 of the 8 equal pieces are shaded. Determine an equivalent fraction if there are 24 pieces. Draw and shade a picture to represent the equivalent fraction.

Extended Response

13. Lori received the following scores on her math quizzes.

    $\frac{10}{15}, \frac{21}{30}, \frac{5}{5}, \frac{56}{60}, \frac{14}{15}$

    a. Write an equivalent fraction for each score using a denominator of 60.

    b. Order the scores from least to greatest. Show your work.

    c. Lori has another quiz tomorrow worth 30 points. If she wants to score the same on this quiz as she did on the one worth 60 points, how many points does she need? Show your work.

Holt Middle School Math   Course 1

Name _____ Date _____ Class _____

Standardized Test Practice Answer Sheet

Chapter 4

Multiple Choice

1. (A) (B) (C) (D)   See Lesson 4-1.
2. (F) (G) (H) (I)   See Lesson 4-6.
3. (A) (B) (C) (D)   See Lesson 4-4.
4. (F) (G) (H) (I)   See Lesson 4-1.

5. (A) (B) (C) (D)   See Lesson 4-8.
6. (F) (G) (H) (I)   See Lesson 4-9.
7. (A) (B) (C) (D)   See Lesson 4-7.

Gridded Response

8.

See Lesson 4-5.

9.

See Lesson 4-3.

10.

See Lesson 4-2.

Short Response

Write your answers in the space provided.

11. _____

_____

_____ (See Lesson 4-3.)

12. _____

_____ (See Lesson 4-4.)

Extended Response

Write your answers for Problem 13 on the back of this paper.

See Lesson 4-6.

Holt Middle School Math   Course 1

Name _____ Date _____ Class _____

Standardized Test Practice Answer Sheet

Chapter 4, continued

13. Part A. _____

_____

_____

_____

_____

Part B. _____

_____

Part C. _____

_____

_____

_____

_____

68   **Holt Middle School Math   Course 1**

Name _____ Date _____ Class _____

Standardized Test Practice Answer Sheet
Chapter 4

Multiple Choice

1. (A) (B) (C) ●   See Lesson 4-1.   5. (A) (B) ● (D)   See Lesson 4-8.

2. (F) (G) ● (I)   See Lesson 4-6.   6. (F) (G) ● (I)   See Lesson 4-9.

3. ● (B) (C) (D)   See Lesson 4-4.   7. (A) (B) (C) ●   See Lesson 4-7.

4. (F) ● (H) (I)   See Lesson 4-1.

Gridded Response

8.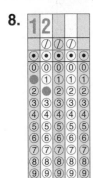

9.

10.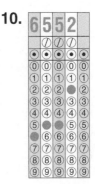

See Lesson 4-5.   See Lesson 4-3.   See Lesson 4-2.

Short Response
Write your answers in the space provided.

11. 3 baskets; To find the greatest equal number of each color, I found the greatest common factor of 12, 16, and 24, which is 4. Because she has 12 green votives, if 4 go into each basket, she can make 3 baskets.   (See Lesson 4-3.)

12. The equivalent fraction is $\frac{18}{24}$.

                                                      (See Lesson 4-4.)

Extended Response
Write your answers for Problem 13 on the back of this paper.
See Lesson 4-6.

Standardized Test Practice Answer Sheet

Chapter 4, continued

13. Part A. $\dfrac{10}{15} = \dfrac{40}{60}$

$\dfrac{21}{30} = \dfrac{42}{60}$

$\dfrac{5}{5} = \dfrac{60}{60}$

$\dfrac{56}{60} = \dfrac{56}{60}$

$\dfrac{14}{15} = \dfrac{56}{60}$

Part B. $\dfrac{40}{60}, \dfrac{42}{60}, \dfrac{56}{60}, \dfrac{56}{60}, \dfrac{60}{60}$

$\dfrac{10}{15}, \dfrac{21}{30}, \dfrac{56}{60}$ or $\dfrac{14}{15}, \dfrac{5}{5}$

Part C. $\dfrac{56}{60} = \dfrac{?}{30}$

$60 = 30 \times 2$

$56 = ? \times 2; \ 56 = 28 \times 2$

$\dfrac{56}{60} = \dfrac{28}{30}$

Lori needs to score a 28 out of 30 on tomorrow's quiz.

Scoring Rubric

4   The student correctly writes the equivalent fractions in the correct order and determines Lori's score. All work is shown.

3   The student correctly writes the equivalent fractions in the correct order, but has minor calculation errors in determining Lori's score. All work is shown.

2   The student correctly writes the equivalent fractions but does not write the fractions in the correct order. There are minor calculation errors in determining Lori's score. Some work is shown.

1   The student correctly writes some of the equivalent fractions but does not write the fractions in the correct order and does not attempt to determine Lori's score. Some work is shown.

0   The answers are not correct and no work is shown.

Holt Middle School Math   Course 1

Name _____ Date _____ Class _____

Standardized Test Practice

Chapter 5

Select the best answer for Questions 1–8.

1. Elkton Jr. High has 36 members in the choir. One-fourth of the members are boys. How many choir members are boys?

   A 4 people

   B 5 people

   C 9 people

   D 12 people

2. Marissa had $\frac{9}{11}$ yard of rope. After she cut some off, she had $\frac{2}{3}$ yard left. How much rope did Marissa cut?

   F $\frac{5}{33}$ yard

   G $\frac{7}{8}$ yard

   H $\frac{18}{33}$ yard

   I $\frac{5}{24}$ yard

3. Taylor walks her neighbor's dog every 5 days and waters the plants every 7 days. How often does Taylor walk the dog and water the plants on the same day?

   A every 5 days

   B every 7 days

   C every 12 days

   D every 35 days

4. A local corporation is entering a team of four to run the Glass City Marathon $(26\frac{1}{5}$ miles). If each runner is to run the same distance, how far will each run?

   F $5\frac{11}{20}$ miles

   G 6 miles

   H $6\frac{11}{20}$ miles

   I $22\frac{1}{5}$ miles

5. A recipe for Chicken Cheese Casserole calls for $\frac{3}{4}$ of a pound of cheese. If you plan to double the recipe, how much cheese will you need?

   A $\frac{3}{4}$ pound

   B 1 pound

   C $1\frac{1}{2}$ pounds

   D 2 pounds

6. Toby is sewing a wedding dress. He has purchased $10\frac{5}{8}$ yd of white satin. Toby cuts off $6\frac{3}{4}$ yd of the satin. How many yards remain?

   F $3\frac{1}{8}$ yd

   G $3\frac{7}{8}$ yd

   H $4\frac{1}{8}$ yd

   I $17\frac{3}{8}$ yd

7. The Mississippi River, in St. Louis had a record flood stage of $43\frac{1}{5}$ feet. In 1993 a record high was set of $49\frac{1}{2}$ feet. How many feet higher was the new record?

   A $6\frac{3}{10}$ ft

   B $6\frac{1}{10}$ ft

   C $5\frac{9}{10}$ ft

   D $6\frac{1}{2}$ ft

8. Maria made 16 cups of juice. She wants to divide it into equal servings. If each serving is $\frac{3}{4}$ cup, which equation could be used to find how many servings she has?

   F $\frac{3}{4}s = 16$

   G $16s = \frac{3}{4}$

   H $\frac{s}{16} = \frac{3}{4}$

   I $s + \frac{3}{4} = 16$

Holt Middle School Math   Course 1

Standardized Test Practice

Chapter 5, continued

Gridded Response

Solve the problems. Use the answer sheet to write and grid-in your answer.

9. Luis and his brother, Rudy, run laps together at the local track. Luis runs a lap in 4 minutes and Rudy runs a lap in 6 minutes. If they start at the same time, how many minutes will it be before they meet again at the starting point?

10. What whole number is the best **estimate** for how far Bob cycled during the week?

Bob's Daily Cycling Mileage	
July 26	$15\frac{1}{2}$
July 27	$22\frac{1}{4}$
July 28	$21\frac{7}{8}$
July 29	$18\frac{1}{3}$
July 30	$16\frac{2}{3}$

11. Gail is making a bookcase for her grandmother's books. Each shelf should be $3\frac{2}{3}$ feet long. If she has to cut $2\frac{1}{3}$ feet from each board, how long are the boards she started with?

Short Response

Solve the problems. Use the answer sheet to write your answers.

12. Fire fighters must reach any fire in the city within $7\frac{3}{4}$ minutes. They are currently reaching fires in $3\frac{1}{2}$ minutes less than the maximum. Write and solve an equation to show how much time it takes them to reach a fire.

13. What is the perimeter of the figure shown? Show all your work.

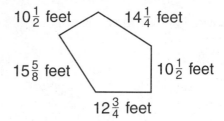

$10\frac{1}{2}$ feet $14\frac{1}{4}$ feet
$15\frac{5}{8}$ feet $10\frac{1}{2}$ feet
$12\frac{3}{4}$ feet

Extended Response

14. Scissors are sold in packs of 5, and thread is sold in packs of 4. Julia wants to give each of her 40 seamstresses one pair of scissors and one spool of thread.

a. Explain in words how to determine the least number of packs of scissors and thread she should buy so there are no extras left over.

b. What is the least number of packs of each she should buy?

c. Julia also wants to give each seamstress a thimble. What is the least number of packs she should buy of thimbles, if thimbles come with 10 in a pack? Show your work.

Holt Middle School Math   Course 1

Name _____ Date _____ Class _____

Standardized Test Practice Answer Sheet

Chapter 5

Multiple Choice

1. Ⓐ Ⓑ Ⓒ Ⓓ   See Lesson 5-1.
2. Ⓕ Ⓖ Ⓗ Ⓘ   See Lesson 5-7.
3. Ⓐ Ⓑ Ⓒ Ⓓ   See Lesson 5-5.
4. Ⓕ Ⓖ Ⓗ Ⓘ   See Lesson 5-3.

5. Ⓐ Ⓑ Ⓒ Ⓓ   See Lesson 5-1.
6. Ⓕ Ⓖ Ⓗ Ⓘ   See Lesson 5-9.
7. Ⓐ Ⓑ Ⓒ Ⓓ   See Lesson 5-9.
8. Ⓕ Ⓖ Ⓗ Ⓘ   See Lesson 5-4.

Gridded Response

9.

See Lesson 5-5.

10.

See Lesson 5-6.

11.

See Lesson 5-8.

Short Response

Write your answers in the space provided.

12. _____

_____

_____
(See Lesson 5-10.)

13. _____

_____

_____
(See Lesson 5-8.)

Extended Response

Write your answers for Problem 14 on the back of this paper.
See Lesson 5-5.

Holt Middle School Math   Course 1

Name _____ Date _____ Class _____

Standardized Test Practice Answer Sheet

Chapter 5, continued

14. Part a. _____

Part b. _____

_____

_____

_____

_____

_____

Part c. _____

_____

_____

_____

Holt Middle School Math   Course 1

Name _____ Date _____ Class _____

Standardized Test Practice Answer Sheet
Chapter 5

Multiple Choice

1. (A) (B) ⬤ (D) See Lesson 5-1.
2. ⬤ (G) (H) (I) See Lesson 5-7.
3. (A) (B) (C) ⬤ See Lesson 5-5.
4. (F) (G) ⬤ (I) See Lesson 5-3.

5. (A) (B) ⬤ (D) See Lesson 5-1.
6. (F) ⬤ (H) (I) See Lesson 5-9.
7. ⬤ (B) (C) (D) See Lesson 5-9.
8. ⬤ (G) (H) (I) See Lesson 5-4.

Gridded Response

9.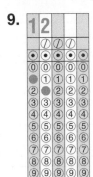

See Lesson 5-5.

10.

See Lesson 5-6.

11.

See Lesson 5-8.

Short Response
Write your answers in the space provided.

12. $t + 3\frac{1}{2} = 7\frac{3}{4}$; $t = 7\frac{3}{4} - 3\frac{1}{2}$; $t = 4\frac{1}{4}$; Currently they are reaching fires in 4 and one-quarter minutes.

(See Lesson 5-10.)

13. $10\frac{1}{2} + 14\frac{1}{4} + 15\frac{5}{8} + 12\frac{3}{4} + 10\frac{1}{2} = 10\frac{4}{8} + 14\frac{2}{8} + 15\frac{5}{8} + 12\frac{6}{8} + 10\frac{4}{8} = 61\frac{21}{8} = 63\frac{5}{8}$; $63\frac{5}{8}$ feet

(See Lesson 5-8.)

Extended Response
Write your answers for Problem 14 on the back of this paper.
See Lesson 5-5.

Holt Middle School Math   Course 1

Standardized Test Practice Answer Sheet
Chapter 5, continued

14. Part a. <u>Find the factors with 5 and 4 that equal 40.</u>

Part b. <u>8 packs of scissors times 5 equal 40 and 10 packs of thread</u>
<u>times 4 equal 40; So, if Julia buys 8 packs of scissors and</u>
<u>10 packs of thread, she can give one of each to each of her</u>
<u>40 seamstress without any left over.</u>

Part c. <u>4 packs of thimbles times 10 equals 40; So, if Julia buys</u>
<u>4 packs of thimbles, she can give one to each of her</u>
<u>40 seamstress without any left over.</u>

Scoring Rubric

4   The student correctly explains the process and correctly determines the number of thimble and scissor packs Julia needs. All work is shown.

3   The student correctly explains the process and correctly determines the number of scissor and thimble packs Julia needs, but does not show all of the work.

2   The student correctly explains the process but has minor calculation errors in determining the correct number of scissor and/or thimble packs Julia needs. Some work is shown.

1   The student incompletely explains the process and has minor calculation errors in determining the correct number of scissor and/or thimble packs Julia needs. Little to no work is shown.

0   The answers are not correct and no work is shown.

Holt Middle School Math   Course 1

Name _____ Date _____ Class _____

Standardized Test Practice

Chapter 6

Select the best answer for Questions 1–5.

1. If you add 5 to every item in a data set what will happen to the mean?

   A It is multiplied by 5.

   B It is increased by 5.

   C It is divided by 5.

   D The mean does not change.

Use the bar graph for Questions 2 and 3.

American Kennel Club Registration

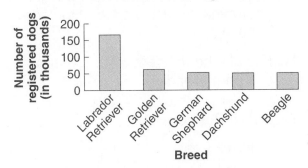

2. A flea collar company is planning a new advertisement campaign. They plan to use the most popular breed of puppy in their ad. Which breed of puppy should they use?

   F Beagle

   G German Shephard

   H Golden Retriever

   I Labrador Retriever

3. What was the number of registered golden retrievers in 2001?

   A 40,000

   B 60,000

   C 50,000

   D 160,000

4. What are the coordinates of point *A*?

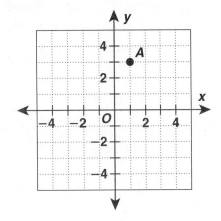

   F (−1, 3)

   G (3, 1)

   H (3, −1)

   I (1, 3)

5. What makes this bar graph misleading?

Sales Profits

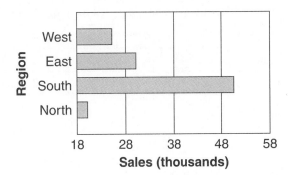

   A The title is incorrect.

   B The horizontal scale does not start at zero.

   C The vertical scale does not start at zero.

   D The space between the data values is not the same.

Holt Middle School Math   Course 1

Standardized Test Practice

Chapter 6, continued

Gridded Response

Solve the problems. Use the answer sheet to write and grid-in your answer.

6. Identify the outlier of the following set of test scores.

60, 64, 59, 63, 88, 67, 70

7. How many servings will you get from 6 cartons of rainbow sherbet?

Sherbet Punch	
Cartons of Sherbet	Number of Servings
2	8
3	12
4	16
5	?
6	?

8. How many points total did Kayla have for all three volleyball games?

	Martin	Keegan	Kayla	Madison
Game 1	II	III	I	IIII
Game 2	IIIII	II	I	II
Game 3	II	III	IIII	III

Short Response

Solve the problems. Use the answer sheet to write your answers.

9. Use the given data to construct a line graph. Be sure to include all parts of the graph.

Ticket Prices at the Zoo	
1992	$3.75
1994	$5.00
1996	$5.00
1998	$6.00
2000	$6.50
2002	$7.00

10. Explain how to use the table in Question 9 to determine the ticket price at the zoo in 2004.

11. Explain the different ways a graph can be misleading.

Extended Response

12.

Stem	Leaf
2	1 1 2 3 4 4
3	1 2
4	0 0 5 5 6
5	7 8 9
6	1 3 3 4 4 5

Key: 2 | 3 means 23

a. What is the least number in the data set? What is the greatest number? Show how to determine the range of the data.

b. Find the mean, median and mode of the data set. Show your work.

c. Is it easier to find the mean, median, or mode from a stem-and-leaf plot? Explain why.

Holt Middle School Math   Course 1

Name _____ Date _____ Class _____

Standardized Test Practice Answer Sheet

Chapter 6

Multiple Choice

1. Ⓐ Ⓑ Ⓒ Ⓓ   See Lesson 6-2.   4. Ⓕ Ⓖ Ⓗ Ⓘ   See Lesson 6-6.

2. Ⓕ Ⓖ Ⓗ Ⓘ   See Lesson 6-4.   5. Ⓐ Ⓑ Ⓒ Ⓓ   See Lesson 6-8.

3. Ⓐ Ⓑ Ⓒ Ⓓ   See Lesson 6-4.

Gridded Response

6.

7.

8.

See Lesson 6-3.   See Lesson 6-1.   See Lesson 6-5.

Short Response

Write your answers in the space provided.

9.   See Lesson 6-7.

10. _____

_____

_____ (See Lesson 6-1.)

11. _____

_____ (See Lesson 6-8.)

Extended Response   See Lesson 6-9.

Write your answers for Problem 12 on the back of this paper.

Holt Middle School Math   Course 1

Standardized Test Practice Answer Sheet

Chapter 6, continued

12. Part A. _____

_____

_____

_____

_____

Part B. _____

_____

_____

_____

_____

Part C. _____

_____

_____

Holt Middle School Math   Course 1

Name _____ Date _____ Class _____

Standardized Test Practice Answer Sheet

Chapter 6

Multiple Choice

1. (A) ● (C) (D) See Lesson 6-2.
2. (F) (G) (H) ● See Lesson 6-4.
3. (A) ● (C) (D) See Lesson 6-4.

4. (F) (G) (H) ● See Lesson 6-6.
5. (A) ● (C) (D) See Lesson 6-8.

Gridded Response

6.

See Lesson 6-3.

7.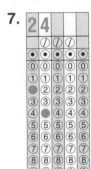

See Lesson 6-1.

8.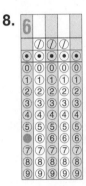

See Lesson 6-5.

Short Response

Write your answers in the space provided.

9.

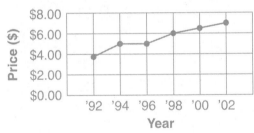

See Lesson 6-7.

10. The table shows that each year the price of a ticket at the zoo increases by 50 cents. So the price of a ticket in 2003 will be $7.00 + 0.50 = 7.50.

(See Lesson 6-1.)

11. A graph can be misleading if a broken axis or distorted pictures are used.

(See Lesson 6-8.)

Extended Response

See Lesson 6-9.

Write your answers for Problem 12 on the back of this paper.

Holt Middle School Math   Course 1

Standardized Test Practice Answer Sheet
Chapter 6, continued

12. Part A. The least number is 21 and the greatest number is 65.

The range is the difference between the greatest and the

least number.

Range = 65 − 21 = 44

Part B. mean =

$$\frac{21 + 21 + 22 + 23 + 24 + 24 + 31 + 32 + 40 + 40 + 45 + 45 + 46 + 57 + 58 + 59 + 61 + 63 + 63 + 64 + 64 + 65}{22}$$

= 44

median = 45

mode = 21, 24, 40, 45, 63, 64

Part C. Possible answer: The mode because you can quickly see what

digits repeat.

Scoring Rubric

4 The student shows correct answers for the extremes and central tendencies,
   as well as, provides a complete explanation for part c. All of the work is shown.

3 The student shows correct answers for the extremes and central tendencies,
   as well as, provides a complete explanation for part c. Some of the work is shown.

2 The student shows correct answers for the extremes, but has some incorrect
   answers due to simple miscalculations for the central tendencies. There is an
   incomplete explanation for part c. Some of the work is shown.

1 The student shows correct answers for the extremes, but has some incorrect
   answers due to simple miscalculations for the central tendencies. Little to no
   work is shown and an explanation for part c is not given.

0 The answers are not correct and no work is shown.

Holt Middle School Math **Course 1**

Name _____ Date _____ Class _____

Standardized Test Practice

Chapter 7

Select the best answer for Questions 1–8.

1. Which figure has four congruent sides?

  A hexagon

  B rhombus

  C triangle

  D trapezoid

2. What is the missing length of the congruent triangles?

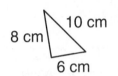 

  F 2 cm

  G 6 cm

  H 8 cm

  I 10 cm

3. The figure shown appears to be an example of what kind of triangle?

  A acute

  B right

  C obtuse

  D equilateral

4. Which letter does NOT have a line of symmetry?

  F H

  G S

  H O

  I W

5. The figure shown is what type of transformation?

  A translation

  B glide reflection

  C reflection

  D rotation

6. What is the name of the shape?

  F quadrilateral

  G pentagon

  H hexagon

  I octagon

7. Which of the following cannot tessellate the plane?

  A hexagon

  B square

  C octagon

  D equilateral triangle

8. Which is a good estimation of the measure of the angle?

  F 16°

  G 25°

  H 58°

  I 131°

Holt Middle School Math   Course 1

Standardized Test Practice

Chapter 7, continued

Gridded Response

Solve the problems. Use the answer sheet to write and grid-in your answer.

9. What is the sum of the angles of a hexagon?

Use the figure to answer Questions 10 and 11.

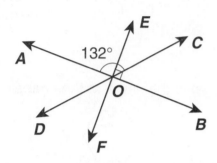

10. What is the measure of ∠BOD?

11. What is the measure of ∠COB?

Short Response

Solve the problems. Use the answer sheet to write your answers.

12. Explain in words the pattern. Then draw the next figure.

13. Compare and contrast perpendicular lines and intersecting lines. Draw an example of each. What symbol do you use to show that two lines are perpendicular?

Extended Response

14. Use the diagram.

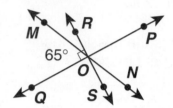

a. Explain in words the relationship between ∠MOQ and ∠PON. Find the measure of ∠PON.

b. Classify ∠ROQ. What is its measure?

c. Explain in words how to find the measure of ∠MOR and ∠NOS.

Holt Middle School Math   Course 1

Standardized Test Practice Answer Sheet

Chapter 7

Multiple Choice

1. Ⓐ Ⓑ Ⓒ Ⓓ   See Lesson 7-6.
2. Ⓕ Ⓖ Ⓗ Ⓘ   See Lesson 7-9.
3. Ⓐ Ⓑ Ⓒ Ⓓ   See Lesson 7-5.
4. Ⓕ Ⓖ Ⓗ Ⓘ   See Lesson 7-11.

5. Ⓐ Ⓑ Ⓒ Ⓓ   See Lesson 7-10.
6. Ⓕ Ⓖ Ⓗ Ⓘ   See Lesson 7-7.
7. Ⓐ Ⓑ Ⓒ Ⓓ   See Lesson 7-12.
8. Ⓕ Ⓖ Ⓗ Ⓘ   See Lesson 7-2.

Gridded Response

9.

See Lesson 7-7.

10.

See Lesson 7-3.

11.

See Lesson 7-3.

Short Response

Write your answers in the space provided.

12. _____

_____

_____

_____ (See Lesson 7-8.)

13. _____

_____

_____ (See Lesson 7-4.)

Extended Response

Write your answers for Problem 14 on the back of this paper.

See Lesson 7-3.

Holt Middle School Math   Course 1

Standardized Test Practice Answer Sheet

Chapter 7, continued

14. Part A. _____

_____

_____

Part B. _____

Part C. _____

_____

_____

_____

Holt Middle School Math   Course 1

Name _____ Date _____ Class _____

Standardized Test Practice Answer Sheet

Chapter 7

Multiple Choice

1. (A) ● (C) (D)   See Lesson 7-6.
2. (F) (G) ● (I)   See Lesson 7-9.
3. (A) (B) ● (D)   See Lesson 7-5.
4. (F) ● (H) (I)   See Lesson 7-11.

5. (A) (B) (C) ●   See Lesson 7-10.
6. (F) (G) ● (I)   See Lesson 7-7.
7. (A) (B) ● (D)   See Lesson 7-12.
8. (F) (G) ● (I)   See Lesson 7-2.

Gridded Response

9.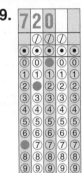

See Lesson 7-7.

10.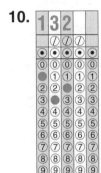

See Lesson 7-3.

11.

See Lesson 7-3.

Short Response
Write your answers in the space provided.

12. The pattern is three shapes: a circle, a square, and a triangle. The first series has the three shapes with no dots. The second series has three shapes with one dot, and the third series has three shapes with two dots, etc. The next shape is a square with two dots.    (See Lesson 7-8.)

13. Perpendicular lines are lines that intersect at a 90° angle. Intersecting lines can intersect at any angle. You use a small square to represent that the two lines are perpendicular.   (See Lesson 7-4.)

Intersecting lines

Perpendicular lines

Extended Response
Write your answers for Problem 14 on the back of this paper.
See Lesson 7-3.

Holt Middle School Math   Course 1

Standardized Test Practice Answer Sheet
Chapter 7, continued

---

14. Part A. $\angle MOQ$ and $\angle PON$ are vertical angles and are therefore congruent. $m\angle PON = m\angle MOQ = 65°$

Part B. $\angle ROQ$ is a right angle. $m\angle ROQ = 90°$

Part C. Since $\angle ROQ$ is a right angle, then $m\angle MOQ + m\angle MOR = 90°$. $m\angle MOQ = 65°$, so $m\angle MOR = 25°$. $\angle MOR$ and $\angle NOS$ are vertical angles and are therefore congruent. So, if $\angle MOR = 25°$ then $m\angle NOS = 25°$.

Scoring Rubric

4   The student recognizes the vertical and right angle relationships, correctly obtains the given angle measurements, and provides a complete explanation. All work is shown.

3   The student recognizes the vertical and right angle relationships, but has some minor miscalculations for the angle measurements. A complete explanation and work are shown.

2   The student correctly classifies one of the angles, but does not recognize one or both angle relationships and has some minor miscalculations for the angle measurements. Some explanation and work are shown.

1   The student correctly classifies the right (90°) angle, but does not recognize any angle relationships. Little work or explanation is shown.

0   The answers are not correct and no work is shown.

Holt Middle School Math   Course 1

Standardized Test Practice
Chapter 8

Select the best answer for Questions 1–7.

1. At Video Mania, 35% of all the rental movies are comedies. What is this percentage as a fraction in simplest form?

   A $\frac{7}{20}$

   B $\frac{17}{50}$

   C $\frac{13}{20}$

   D $\frac{35}{100}$

2. A car salesman has sold 15 vehicles over the past 4 months. If 20% of the vehicles sold have been trucks, how many trucks has the salesman sold during this time?

   F 3 trucks

   G 6 trucks

   H 9 trucks

   I 187 trucks

3. Which is NOT an equivalent ratio to compare the number of suns to moons?

   A $\frac{18}{24}$

   B $\frac{3}{6}$

   C $\frac{9}{12}$

   D $\frac{24}{32}$

4. Your three dogs eat 25 pounds of dog food in 2 weeks. How many pounds will they eat in one year?

   F 300 lb

   G 600 lb

   H 650 lb

   I 1,300 lb

5. What is the missing length and the measure of ∠J in the similar triangles below?

   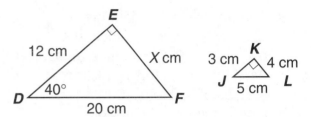

   A 12 cm, 90°

   B 16 cm, 40°

   C 16 cm, 50°

   D 20 cm, 40°

6. A beverage stand at a movie theater sells 4 different sizes of drinks. Which is a better deal?

   F 12 oz for $1.25

   G 16 oz for $1.65

   H 20 oz for $2.15

   I 32 oz for $3.00

7. To make 36 oz of homemade lemonade, you combine 3 cups of cold water with 1 cup of lemon juice and $\frac{1}{2}$ cup of sugar. If you only have enough lemons for a $\frac{1}{2}$ cup of juice, how much sugar will you need?

   A $\frac{1}{8}$ cup

   B $\frac{1}{4}$ cup

   C $\frac{3}{4}$ cup

   D 1 cup

Holt Middle School Math   Course 1

Standardized Test Practice

Chapter 8, continued

Gridded Response

Solve the problems. Use the answer sheet to write and grid-in your answer.

8. A father builds his daughter a dollhouse that is modeled after his own home. The dollhouse measures 24 inches long and 16 inches wide. If the actual house is 40 feet wide, how long is the actual house?

9. A 30 ft building casts a 45 ft shadow and a tree next to the building casts a 24 ft shadow. How tall is the tree?

10. On a map, the distance from your house to the capital building is 4.2 inches. If the scale of the map is 1 inch = 5 miles, how far is the actual distance from your house to the capital building?

Short Response

Solve the problems. Use the answer sheet to write your answers.

11. Explain in words how to write 0.38 as a fraction and a percent.

12. In the 2001, Virginia elections, approximately 46% of the 4,109,127 registered voters actually cast their ballot. How many voters voted? Round to the nearest person. How many registered voters did not cast their ballot?

13. You bought a pair of shoes for $36 and four pairs of socks for $16. You pay a 6% sales tax. How much is the sales tax? How much is the total?

14. Out of 350 6th graders, 73% of the students raised enough money to go to Space Camp. Explain in words how to write this percent as a decimal.

Extended Response

15. During a fundraiser, potential customers had a choice of buying a book cover, a coffee mug with the school logo, or a school pennant. Of the 400 items that were sold, 12% were coffee mugs.

a. Write a proportion to find how many coffee mugs were sold.

b. If 240 book covers were sold, what percent of the items sold were book covers?

c. For next year's sale, the committee has decided to sell only two items. Which two items would you suggest to the committee and why?

Holt Middle School Math   Course 1

Name _____ Date _____ Class _____

Standardized Test Practice Answer Sheet
Chapter 8

Multiple Choice

1. Ⓐ Ⓑ Ⓒ Ⓓ   See Lesson 8-7.   5. Ⓐ Ⓑ Ⓒ Ⓓ   See Lesson 8-4.

2. Ⓕ Ⓖ Ⓗ Ⓘ   See Lesson 8-9.   6. Ⓕ Ⓖ Ⓗ Ⓘ   See Lesson 8-1.

3. Ⓐ Ⓑ Ⓒ Ⓓ   See Lesson 8-1.   7. Ⓐ Ⓑ Ⓒ Ⓓ   See Lesson 8-2.

4. Ⓕ Ⓖ Ⓗ Ⓘ   See Lesson 8-3.

Gridded Response

8.

See Lesson 8-4.

9.

See Lesson 8-5.

10.

See Lesson 8-6.

Short Response
Write your answers in the space provided.

11. _____

_____
(See Lesson 8-8.)

12. _____

_____
(See Lesson 8-10.)

13. _____

_____
(See Lesson 8-9.)

14. _____

_____
(See Lesson 8-4.)

Extended Response
Write your answers for Problem 15 on the back of this paper. See Lesson 8-4.

Holt Middle School Math   Course 1

Name _____ Date _____ Class _____

Standardized Test Practice Answer Sheet
Chapter 8, continued

15. Part A. _____

Part B. _____

Part C. _____

_____

_____

_____

Holt Middle School Math   Course 1

Name _____ Date _____ Class _____

Standardized Test Practice Answer Sheet

Chapter 8

Multiple Choice

1. ● Ⓑ Ⓒ Ⓓ   See Lesson 8-7.
2. ● Ⓖ Ⓗ Ⓘ   See Lesson 8-9.
3. Ⓐ ● Ⓒ Ⓓ   See Lesson 8-1.
4. Ⓕ Ⓖ ● Ⓘ   See Lesson 8-3.

5. Ⓐ ● Ⓒ Ⓓ   See Lesson 8-4.
6. Ⓕ Ⓖ Ⓗ ●   See Lesson 8-1.
7. Ⓐ ● Ⓒ Ⓓ   See Lesson 8-2.

Gridded Response

8.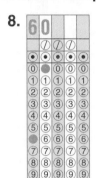

See Lesson 8-4.

9.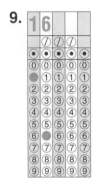

See Lesson 8-5.

10.

See Lesson 8-6.

Short Response
Write your answers in the space provided.

11. To write 0.38 as a percent, multiply the number by 100. 0.38 = 38% To write 0.38 as a fraction, write it over 100 and simplify: $\frac{38}{100} = \frac{19}{50}$. (See Lesson 8-8.)

12. $0.46 \times 4,109,127 = 1,890,198$ people; $4,109,127 - 1890,198 = 2,218,929$ people (See Lesson 8-10.)

13. $36 + 16 = 52$; $52 \times 0.06 = 3.12$; You pay $3.12 in sales tax. $52 + 3.12 = \$55.12$; You pay $55.12 total. (See Lesson 8-9.)

14. Write the percent as a fraction with a denominator of 100. Then divide. $\frac{73}{100} = 73\overline{)100} = 0.73$ (See Lesson 8-4.)

Extended Response
Write your answers for Problem 15 on the back of this paper. See Lesson 8-4.

Holt Middle School Math   Course 1

Standardized Test Practice Answer Sheet

Chapter 8, continued

15. Part A. $\dfrac{c}{400} = \dfrac{12}{100}$; $c = 48$; 48 coffee mugs

Part B. $\dfrac{240}{400} = \dfrac{x}{100}$; $x = 60$; 60% of the sold items were book covers.

Part C. The two items that should be sold next year are the book covers and the pennant. Of the 400 sold items, 352 of them were either book covers or pennants. These seem to be the best selling items. Not enough of the coffee mugs were sold.

Scoring Rubric

4   The student writes a proportion to arrive at the correct solution, determines the correct percentage of book covers sold, and has a viable explanation for the committee.

3   The student writes the correct proportion and has a good explanation for the committee, but has minor errors in calculations.

2   The student writes the correct proportion, but has calculation errors for the solution and does not have a good explanation for the committee.

1   The student does not write any proportions, but correctly answers the problem. Explanations are not correct or not given.

0   The answers are not correct and no work is shown.

Holt Middle School Math   Course 1

Standardized Test Practice

Chapter 9

Select the best answer for Questions 1–8.

1. Evaluate $w \div (-4)$ for $w = -44$.

    A −11   **C** 10

    B −10   **D** 11

2. The temperature outside is −4°C. It drops 6° at night. What is the new temperature?

    F −10°   **G** −2°

    H 2°   **I** 10°

3. Mrs. Lee kept a record of the changes in the water level as the tide changed. At what time did the water level increase the most?

Time	Change in Water Level (inches)
6:00 A.M.	−6
8:00 A.M.	−4
10:00 A.M.	0
noon	+3
2:00 P.M.	+6
4:00 P.M.	+11

    A between 6:00 A.M. and 8:00 A.M.

    B between 8:00 A.M. and 10:00 A.M.

    C between noon and 2:00 P.M.

    D between 2:00 P.M. and 4:00 P.M.

4. At the Fall Festival, Mr. Moulter's class had expenses of $5, $2, and $3. The sales at the end of the festival were $6, $8, and $3. What was their total profit?

    F −$7   **H** $10

    G $7   **I** $17

5. Chelsea and 3 other friends went golfing. Use the table to determine who won. (Hint: Low score wins.)

Player	Score
Chelsea	−2
Bryan	−5
Laurie	2
Todd	5

    A Chelsea

    B Bryan

    C Laurie

    D Todd

6. On a coordinate plane, the point where the axes intersect is called the _____.

    F quadrant

    G origin

    H ordered pair

    I coordinate point

7. In a board game, Trina made the following moves: 10 spaces forward, 3 spaces back, 2 spaces forward, and 5 spaces back. Where is Trina's position from where she started?

    A back 4 spaces

    B back 8 spaces

    C forward 4 spaces

    D forward 6 spaces

8. What number must 12 be multiplied by to get a product of −12?

    F 0   **H** −1

    G 1   **I** −12

Holt Middle School Math   Course 1

Name _____ Date _____ Class _____

Standardized Test Practice
Chapter 9, continued

Gridded Response
Solve the problems. Use the answer sheet to write and grid-in your answer.

9. In one week the temperature rose 16° to reach 10°F. The temperature was originally negative ____°.

10. You spent $25 on groceries and $31 at the bookstore. You had $150 before you began shopping. How much do you have left?

Short Response
Solve the problems. Use the answer sheet to write your answers.

11. Graph and label each point on the coordinate plane.

A (3, 4)

B (−2, −4)

C (0, −3)

D (2, 0)

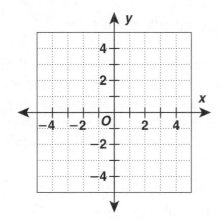

12. A submarine starts at sea level and dives 500 feet, rises 50 feet, but then has to rise 175 feet more to reach periscope depth. After receiving a report from the base, the submarine dives by an amount equal to 3 times its first rise. Write an expression to represent the submarine's change in sea level and then find the final depth of the submarine.

13. Explain how positive and negative integers represent transactions with money.

Extended Response

14. Larry has been investing his money in the stock market. He bought 25 shares of a mutual fund for $530. He sold the fund for a profit of $120.

a. Write and solve an equation to find the price per share at which Larry sold his fund.

b. The price of the fund dropped $80 the day after Larry sold it. Write and solve an equation to find the price Larry would have sold it if he had waited until the next day.

c. Would Larry have earned a profit if he had waited until the next day to sell the stock? Explain.

Holt Middle School Math   Course 1

Name _____ Date _____ Class _____

Standardized Test Practice Answer Sheet
Chapter 9

Multiple Choice

1. Ⓐ Ⓑ Ⓒ Ⓓ See Lesson 9-7.
2. Ⓕ Ⓖ Ⓗ Ⓘ See Lesson 9-5.
3. Ⓐ Ⓑ Ⓒ Ⓓ See Lesson 9-4.
4. Ⓕ Ⓖ Ⓗ Ⓘ See Lesson 9-4.

5. Ⓐ Ⓑ Ⓒ Ⓓ See Lesson 9-2.
6. Ⓕ Ⓖ Ⓗ Ⓘ See Lesson 9-3.
7. Ⓐ Ⓑ Ⓒ Ⓓ See Lesson 9-4.
8. Ⓕ Ⓖ Ⓗ Ⓘ See Lesson 9-6.

Gridded Response

9.

10.

See Lesson 9-5.   See Lesson 9-8.

Short Response

Write your answers in the space provided.

11.

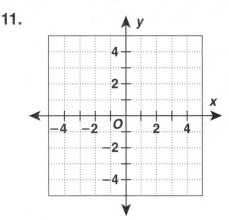

(See Lesson 9-3.)

12. _____
_____
_____ (See Lesson 9-6.)

13. _____
_____
_____ (See Lesson 9-1.)

Extended Response

Write your answers for Problem 14 on the back of this paper.

See Lesson 9-8.

Holt Middle School Math   Course 1

Name _____ Date _____ Class _____

Standardized Test Practice Answer Sheet

Chapter 9, continued

14. Part A. _____

_____

Part B. _____

_____

Part C. _____

_____

_____

Holt Middle School Math   Course 1

Name _____ Date _____ Class _____

Standardized Test Practice Answer Sheet
Chapter 9

Multiple Choice

1. (A) (B) (C) ●   See Lesson 9-7.

2. ● (G) (H) (I)   See Lesson 9-5.

3. (A) (B) (C) ●   See Lesson 9-4.

4. (F) ● (H) (I)   See Lesson 9-4.

5. (A) ● (C) (D)   See Lesson 9-2.

6. (F) ● (H) (I)   See Lesson 9-3.

7. (A) (B) ● (D)   See Lesson 9-4.

8. (F) (G) ● (I)   See Lesson 9-6.

Gridded Response

9.

See Lesson 9-5.

10.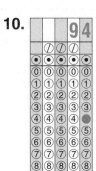

See Lesson 9-8.

Short Response
Write your answers in the space provided.

11.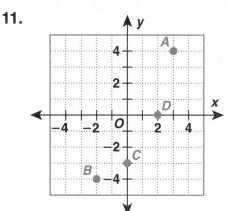

(See Lesson 9-3.)

12. $0 - 500 + 50 + 175 - 3(50)$; The final depth of the submarine is -425 ft.

_____

_____

(See Lesson 9-6.)

13. Positive numbers represent money saved or earned and negative numbers represent money spent.

_____

(See Lesson 9-1.)

Extended Response
Write your answers for Problem 14 on the back of this paper.
See Lesson 9-8.

Holt Middle School Math   Course 1

Standardized Test Practice Answer Sheet
Chapter 9, continued

14. Part A. $25x = 530 + 120$; $25x = 650$; $x = 26$; He sold for $26 per share.

Part B. $25x = 530 + 120 - 80$; $25x = 570$; $x = \$22.80$; He would have sold it for $22.80 per share.

Part C. Yes, he would have made a profit. He would have received $570, and he paid $530 for the fund. So he would have gained $40.

Scoring Rubric

4   The student correctly writes and solves the equations and provides a good explanation for part c.

3   The student correctly writes the equations, but has minor calculation errors in solving them. The student has a good explanation to part c that is consistent with the incorrect solutions to the equations.

2   The student did not write any equations but has correct solutions based on other shown methods. The explanation for part c is complete.

1   The student incorrectly writes one or both equations and has calculation errors in solving them. The explanation for part c is incomplete.

0   The answers are not correct and no explanation is given.

Holt Middle School Math   Course 1

Name _____ Date _____ Class _____

Standardized Test Practice
Chapter 10

Select the best answer for Questions 1–8.

1. Identify the figure shown.

   A square pyramid

   B cube

   C cylinder

   D cone

2. What is the volume of a 5 feet long, 3 feet wide, and 2 feet deep rectangular toy chest?

   F 10 ft^3 **H** 30 ft^3

   G 15 ft^3 **I** 60 ft^3

3. Find the volume of the cylinder. Use 3.14 for π.

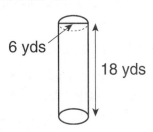

   A 508.7 yd^3 **C** 169.6 yd^3

   B 339.1 yd^3 **D** 162 yd^3

4. What is the volume of the figure shown?

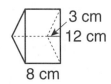

   F 288 cm^3 **H** 48 cm^3

   G 144 cm^3 **I** 12 cm^3

5. What is the perimeter of the polygon?

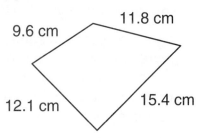

   A 50 cm   **C** 42 cm

   B 48.9 cm   **D** 37.1 cm

6. Mrs. Albright has a rug for her classroom floor. The rug is 20 feet long. What else must she know to calculate how many square feet the rug will occupy?

   F The number of rugs in the room.

   G The width of the classroom.

   H The width of the rug.

   I The height of the classroom.

7. If square floor tile costs $2 per square foot, what is the cost of tiling a rectangular floor 12 feet by 15 feet?

   A $360   **C** $108

   B $180   **D** $54

8. Estimate the area of the figure shown.

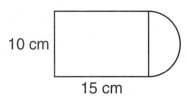

   F 150 cm^2 **H** 189.25 cm^2

   G 175 cm^2 **I** 228.5 cm^2

Holt Middle School Math   **Course 1**

Standardized Test Practice

Chapter 10, continued

Gridded Response

Solve the problems. Use the answer sheet to write and grid-in your answer.

9. Kara wants to cover the surface of the pyramid shown with brown paper for a social studies project. How many square centimeters of paper will she need?

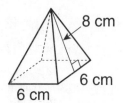

8 cm

6 cm

6 cm

10. What is the area in square inches that a ceiling fan covers if the length of one of the fan blades extends 15 in. from the center? Use 3.14 for π and round to the nearest tenth of an inch.

11. The circumference at the base of the Tower of Pisa is 48.6 m. What is the radius in meters of the Tower of Pisa? Use 3.14 for π.

Short Response

Solve the problems. Use the answer sheet to write your answers.

12. The top of a shed has the shape of a square pyramid. The base edges are 12 ft and the slant height is 8 feet. How much plastic is needed to cover the top of the shed? (Hint: the base of the pyramid is not included.)

13. An architect wants to change the dimensions of a rectangular room he had drafted. In the new draft, each dimension is twice as large as in the first drawing. What happens to the area of the second room in relation to the first room? Find the area of each room.

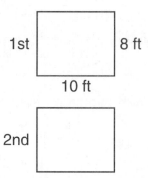

1st

8 ft

10 ft

2nd

Extended Response

14. The state of Kansas is rectangular in shape. It is approximately 401 mi long and 204 mi wide.

   a. Find the area of Kansas.

   b. In 2000, the census calculated that there were 2,688,418 people living in Kansas. Estimate the population per square mile.

   c. If you can cycle 121 miles per day, how many days would it take you to cycle the borders of Kansas? Explain your answer.

Holt Middle School Math   Course 1

Name _____ Date _____ Class _____

Standardized Test Practice Answer Sheet

Chapter 10

Multiple Choice

1. Ⓐ Ⓑ Ⓒ Ⓓ   See Lesson 10-6.
2. Ⓕ Ⓖ Ⓗ Ⓘ   See Lesson 10-8.
3. Ⓐ Ⓑ Ⓒ Ⓓ   See Lesson 10-9.
4. Ⓕ Ⓖ Ⓗ Ⓘ   See Lesson 10-8.

5. Ⓐ Ⓑ Ⓒ Ⓓ   See Lesson 10-1.
6. Ⓕ Ⓖ Ⓗ Ⓘ   See Lesson 10-2.
7. Ⓐ Ⓑ Ⓒ Ⓓ   See Lesson 10-2.
8. Ⓕ Ⓖ Ⓗ Ⓘ   See Lesson 10-3.

Gridded Response

9.

See Lesson 10-7.

10.

See Lesson 10-5.

11.

See Lesson 10-5.

Short Response

Write your answers in the space provided.

12. _____

_____

_____ (See Lesson 10-7.)

13. _____

_____

_____ (See Lesson 10-4.)

Extended Response

Write your answers for Problem 14 on the back of this paper.

See Lesson 10-2.

Holt Middle School Math   Course 1

Standardized Test Practice Answer Sheet
Chapter 10, continued

14. Part A. _____

Part B. _____

Part C. _____

_____

_____

Name _____ Date _____ Class _____

Standardized Test Practice Answer Sheet

Chapter 10

Multiple Choice

1. Ⓐ Ⓑ ⬤ Ⓓ See Lesson 10-6.
2. Ⓕ Ⓖ ⬤ Ⓘ See Lesson 10-8.
3. ⬤ Ⓑ Ⓒ Ⓓ See Lesson 10-9.
4. Ⓕ ⬤ Ⓗ Ⓘ See Lesson 10-8.

5. Ⓐ ⬤ Ⓒ Ⓓ See Lesson 10-1.
6. Ⓕ Ⓖ ⬤ Ⓘ See Lesson 10-2.
7. ⬤ Ⓑ Ⓒ Ⓓ See Lesson 10-2.
8. Ⓕ Ⓖ ⬤ Ⓘ See Lesson 10-3.

Gridded Response

9.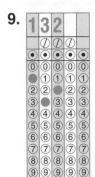

See Lesson 10-7.

10.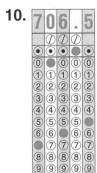

See Lesson 10-5.

11.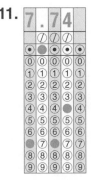

See Lesson 10-5.

Short Response

Write your answers in the space provided.

12. $4\left(\dfrac{1}{2} \cdot 12 \cdot 8\right) = 192$; 192 square feet of plastic is needed.

_____

_____ (See Lesson 10-7.)

13. The area of the second room is four times as large. Area of the first room is 80 sq ft. Area of the second room is 320 sq ft.

_____ (See Lesson 10-4.)

Extended Response

Write your answers for Problem 14 on the back of this paper.
See Lesson 10-2.

Holt Middle School Math   Course 1

Name _____ Date _____ Class _____

Standardized Test Practice Answer Sheet
Chapter 10, continued

---

14. Part A. $A = (401 \text{ mi})(204 \text{ mi}) = 81{,}804 \text{ mi}^2$

Part B. $2{,}688{,}418 \text{ people} \div 81{,}804 \text{ mi}^2 \approx 33 \text{ people/square mile}$

Part C. $P = 2l + 2w$; $p = 2(401) + 2(204) = 1{,}210 \text{ mi}$. The distance around Kansas is 1,210 mi. If you cycle 121 miles per day, you can cycle the borders in $1{,}210 \text{ mi} \div 121 \text{ mi/day} = 10 \text{ days}$.

Scoring Rubric

4   The student correctly answers each section, provides a clear explanation, and shows all the work.

3   The student provides a clear explanation and shows all the work, but has minor calculation errors in some or all of the sections.

2   The student shows some work, but the work contains calculation errors. The explanation is consistent with the incorrect calculations.

1   The student has major miscalculations and the explanation is incomplete. Little to no work is shown.

0   The answers are not correct, no work is shown, and no explanation is given.

Holt Middle School Math   Course 1

Name _____ Date _____ Class _____

Standardized Test Practice

Chapter 11

Select the best answer for Questions 1–7.

1. Find the probability of getting at least 2 tails when a coin is tossed three times.

 A $\frac{1}{4}$ **C** $\frac{1}{2}$

 B $\frac{3}{8}$ **D** $\frac{5}{8}$

2. An antique dealer predicts that 40% of its sales are from antique linens. Predict how much money the dealer has made if she sold $500 worth of linens.

 F $40

 G $100

 H $1,250

 I $1,300

3. Veronica spun the arrow on a spinner 60 times. The results are shown in the table. Which of these spinners did Veronica most likely spin?

Shape	Circle	Triangle	Square	Total Spins
Number of Times	22	20	18	60

 A   **C**

 B   **D**

4. There are 30 jellybeans in a bag. Five are red, 6 yellow, 5 black, 3 orange, 4 green, and 7 purple. If Leona selects one jellybean from the bag, what is the probability that it will be black?

 F 5

 G $\frac{1}{5}$

 H $\frac{5}{30}$

 I 30%

5. You roll a fair number cube 36 times. How many times would you expect to roll a number that is five or greater?

 A 3   **C** 12

 B 6   **D** 18

6. Which describes the chances that Mr. Fredrickson will be late for his 4:00 meeting if he stuck in traffic 30 miles away and it is 3:50?

 F impossible

 G unlikely

 H likely

 I certain

7. A student randomly guessed the answers to three questions on a true-false test. What is the probability that the student correctly answered all three questions?

 A $\frac{1}{8}$ **C** $\frac{1}{3}$

 B $\frac{1}{6}$ **D** $\frac{1}{2}$

Holt Middle School Math   Course 1

Standardized Test Practice
Chapter 11, continued

Gridded Response
Solve the problems. Use the answer sheet to write and grid-in your answer.

8. If there is a 28% chance of snow today, what is the percent chance that it will NOT snow?

9. Gail flips a fair coin and spins a fair spinner numbered from 1 to 6. What is the probability of landing on a number divisible by 3 and the coin showing heads?

10. An artist chooses 3 colors from 6 standard colors for a mural. How many different combinations of colors can the artist chose from?

Short Response
Solve the problems. Use the answer sheet to write your answers.

11. Firefighters can enter a burning building through one of four entrances. They can exit the building through any one of the six exits. How many ways is it possible for them to enter and exit the building? Draw a tree diagram to illustrate how you determined your answer.

12. Katie is running late for work and she still has to put on her shoes and socks. She has 10 pairs of shoes in her closet, of which 1 pair are loafers, 4 pairs are tennis shoes, 2 pairs are sandals, and 3 pairs are high-heels. If Katie randomly picks 1 tennis shoe from the closet, what is the probability that Katie will randomly pick out another tennis shoe? Show your work.

13. What is the probability that a week of the year picked at random will have 7 days? Explain in words how you determined your answer.

Extended Response

14. When a customer enters the Vet Pet Store, an employee randomly hands the customer a pet to hold. Today, the store has 8 puppies, 8 kittens, 12 birds, 5 lizards, and 2 bunnies for sale.

   a. What is the probability that an employee will hand a puppy to a customer?

   b. It has been the manager's experience that an employee gives a customer a lizard to hold 1 out of 10 times. How does this compare with the theoretical probability of the employee choosing a lizard?

   c. Compare the likelihood of a customer receiving a kitten to hold with the likelihood of a customer receiving each of the other animals.

Name _____ Date _____ Class _____

Standardized Test Practice Answer Sheet

Chapter 11

Multiple Choice

1. (A) (B) (C) (D)   See Lesson 11-5.
2. (F) (G) (H) (I)   See Lesson 11-6.
3. (A) (B) (C) (D)   See Lesson 11-2.
4. (F) (G) (H) (I)   See Lesson 11-3.

5. (A) (B) (C) (D)   See Lesson 11-6.
6. (F) (G) (H) (I)   See Lesson 11-1.
7. (A) (B) (C) (D)   See Lesson 11-4.

Gridded Response

8.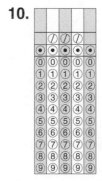

See Lesson 11-3.

9.

See Lesson 11-5.

10.

See Lesson 11-4.

Short Response
Write your answers in the space provided.

11. _____ (See Lesson 11-4.)

12. _____

_____ (See Lesson 11-3.)

13. _____

_____

_____ (See Lesson 11-1.)

Extended Response
Write your answers for Problem 14
on the back of this paper.
See Lessons 11-2 and 11-3.

Holt Middle School Math   Course 1

Name _____ Date _____ Class _____

Standardized Test Practice Answer Sheet
Chapter 11, continued

14. Part A. _____

Part B. _____

_____

_____

Part C. _____

_____

_____

_____

Holt Middle School Math   **Course 1**

Standardized Test Practice Answer Sheet

Chapter 11

Multiple Choice

1. (A) (B) ● (D)   See Lesson 11-5.
2. (F) (G) ● (I)   See Lesson 11-6.
3. (A) ● (C) (D)   See Lesson 11-2.
4. (F) (G) ● (I)   See Lesson 11-3.

5. (A) (B) ● (D)   See Lesson 11-6.
6. (F) (G) (H) ●   See Lesson 11-1.
7. ● (B) (C) (D)   See Lesson 11-4.

Gridded Response

8.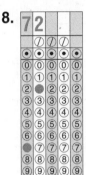

See Lesson 11-3.

9.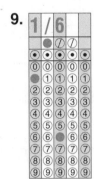

See Lesson 11-5.

10.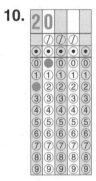

See Lesson 11-4.

Short Response
Write your answers in the space provided.

11. 24 ways   (See Lesson 11-4.)

12. P(choosing a second tennis shoe) $= \dfrac{7}{19}$ (See Lesson 11-3.)

13. The probability is 1 because every week has 7 days.

(See Lesson 11-1.)

Extended Response
Write your answers for Problem 14 on the back of this paper.
See Lessons 11-2 and 11-3.

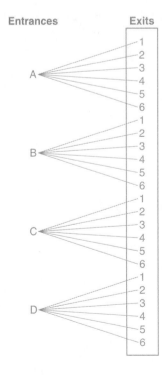

Standardized Test Practice Answer Sheet

Chapter 11, continued

14. Part A. $\dfrac{8}{35}$

Part B. The theoretical probability of the employee handing a customer a lizard is $\dfrac{5}{35} = \dfrac{1}{7}$. This is greater than the experimental probability of $\dfrac{1}{10}$.

Part C. The likelihood of receiving a kitten is the same as the likelihood of receiving a puppy but greater than the likeliness of receiving a lizard or a bunny, and less than the likelihood of receiving a bird.

Scoring Rubric

4 The student shows correct answers for the probabilities and has a clear explanation of the likeliness of the events.

3 The student shows correct answers for the probabilities but the explanation of the likeliness of the events is incomplete.

2 The student has some incorrect answers for the probabilities and has a poor explanation of the likeliness of the events.

1 The student has some incorrect answers for the probabilities and no explanation is given.

0 The answers are not correct and no work is shown.

Name _____ Date _____ Class _____

Standardized Test Practice
Chapter 12

Select the best answer for Questions 1–5.

1. A clothing company charges $12 for each T-shirt and $3 for shipping and handling. Maria paid $75 for some T-shirts. How many T-shirts did Maria order?

   A 5 shirts   **C** 7 shirts

   B 6 shirts   **D** 8 shirts

2. Which graph matches the function $y = 2x + 3$?

F 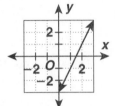   **H**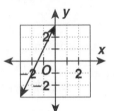

G 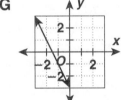   **I**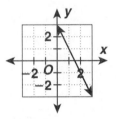

3. Identify the graph of $y = 3$.

A 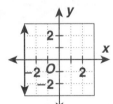   **C**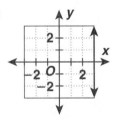

B 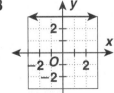   **D**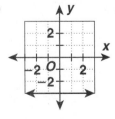

4. Which figure shows a decrease in the horizontal dimensions of $\frac{1}{2}$?

F

G

H

I

5. What are the coordinates of *ABCD* after being translated 2 units left and 2 units up?

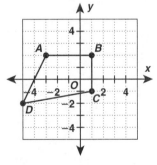

   A (−3, 2); (1, 2); (1, −1); (−5, −2)

   B (−1, 2); (3, 2); (3, −1); (−3, −2)

   C (−1, 0); (3, 0); (3, −3); (−3, −4)

   D (−5, 4); (−1, 4); (−1, 1); (−7, 0)

Holt Middle School Math   **Course 1**

Name _____ Date _____ Class _____

Standardized Test Practice
Chapter 12, continued

Gridded Response
Solve the problems. Use the answer sheet to write and grid-in your answer.

6. The graph of the line $x = -2$ passes through the point $(-?, 6)$.

7. When the point $(-5, -2)$ is reflected across the y-axis, what is the new x-coordinate?

8. When the point $(5, -2)$ is translated four units to the right and 3 units down, the new y-coordinate is $-$____.

Short Response
Solve the problems. Use the answer sheet to write your answers.

9. Graph triangle *MNP* on the coordinate plane then reflect it across the x-axis.

$M(3, -2)$, $N(5, -4)$, and $P(1, -3)$

10. What are the coordinates of *A, B, C,* and *D* after rotating trapezoid *ABCD* 180° about the origin? Plot the rotation on the graph.

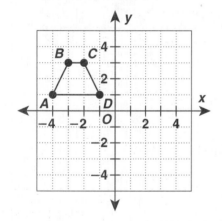

11. A collectible company charges $35 for each figurine sold and a flat fee of $8 for shipping and handling. Write a linear equation to describe this situation, and then determine how many figurines can be purchased with $290?

Extended Response

12. Kim has saved $500 to use for a down payment on a mini-van. She wants to continue saving $50 a month so that her down payment is $1,000.

a. Write a linear equation to model how much money, y, Kim can save in x months.

b. Graph the linear equation.

c. How many months will it take for Kim to save $1,000? Explain how you determined your answer.

Holt Middle School Math   Course 1

Name _____ Date _____ Class _____

Standardized Test Practice Answer Sheet
Chapter 12

Multiple Choice

1. Ⓐ Ⓑ Ⓒ Ⓓ See Lesson 12-1.   4. Ⓕ Ⓖ Ⓗ Ⓘ See Lesson 12-6.

2. Ⓕ Ⓖ Ⓗ Ⓘ See Lesson 12-2.   5. Ⓐ Ⓑ Ⓒ Ⓓ See Lesson 12-3.

3. Ⓐ Ⓑ Ⓒ Ⓓ See Lesson 12-2.

Gridded Response

6.

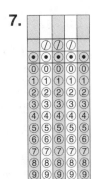

See Lesson 12-2.

7.

See Lesson 12-4.

8.

See Lesson 12-3.

Short Response
Write your answers in the space provided.

9.

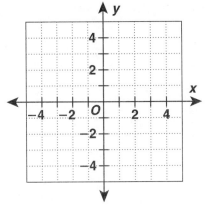

(See Lesson 12-4.)

10.

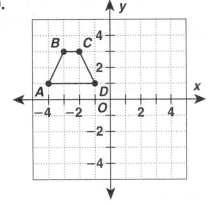

(See Lesson 12-5.)

11. _____

_____ (See Lesson 12-1.)

Extended Response
Write your answers for Problem 12 on the back of this paper.
See Lesson 12-2.

Holt Middle School Math   Course 1

Name _____ Date _____ Class _____

Standardized Test Practice Answer Sheet
Chapter 12, continued

12. Part A. _____

_____

Part B.

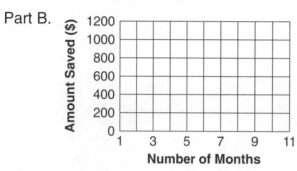

Part C. _____

_____

_____

_____

Holt Middle School Math   Course 1

Name _____ Date _____ Class _____

Standardized Test Practice Answer Sheet

Chapter 12

Multiple Choice

1. (A) ● (C) (D)   See Lesson 12-1.
2. (F) (G) ● (I)   See Lesson 12-2.
3. (A) ● (C) (D)   See Lesson 12-2.

4. (F) ● (H) (I)   See Lesson 12-6.
5. (A) (B) (C) ●   See Lesson 12-3.

Gridded Response

6. 2

See Lesson 12-2.

7. 5

See Lesson 12-4.

8. 5

See Lesson 12-3.

Short Response

Write your answers in the space provided.

9.

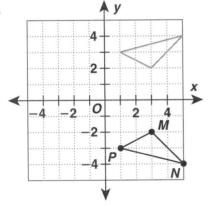

(See Lesson 12-4.)

10.

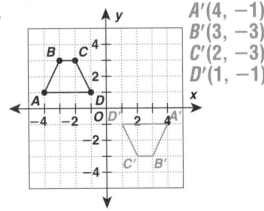

$A'(4, -1)$
$B'(3, -3)$
$C'(2, -3)$
$D'(1, -1)$

(See Lesson 12-5.)

11. $y = 35x + 8$; $290 = 35x + 8$; $282 = 35x$; $8.01 \approx x$; 8 figures can be purchased with $290.

(See Lesson 12-1.)

Extended Response

Write your answers for Problem 12 on the back of this paper.

See Lesson 12-2.

Holt Middle School Math   Course 1

Standardized Test Practice Answer Sheet
Chapter 12, *continued*

12. Part A.   Let *x* represent number of months Kim saves money.

$$y = 500 + 50x$$

Part B.

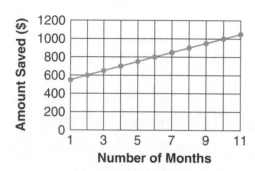

Part C.   Possible answer: I used the graph and followed the line from

$y = 1,000$ to the line. I then read the mark on the *x*-axis:

10 months; or I solve the equation for *x*. $1,000 = 500 + 50x$;

$500 = 50x$; $x = 10$; It will take her 10 months to save $1,000.

Scoring Rubric

4   The student correctly writes an equation, and then correctly graphs and solves the equation. The explanation correctly explains the solution.

3   The student correctly writes an equation, and correctly graphs the equation, but has minor calculation errors is solving the equation. The explanation accurately explains the solution.

2   The student correctly writes the equation but has some errors in the graph and/or the solution. The explanation is incomplete.

1   The student incorrectly writes the equation, and there are some errors in the graph and/or solution. The explanation is incomplete or unclear.

0   The answers are not correct and no work is shown

Holt Middle School Math   Course 1

Name _____ Date _____ Class _____

End-of-Grade Practice Test 1

Directions: Read each problem carefully, and fill in the oval that corresponds to the correct answer.

1. Michael has a sailboat that is 24 ft long. What is the boat's length in yards?
   - Ⓐ 3
   - Ⓑ 8
   - Ⓒ 10
   - Ⓓ 21

2. Fill in the blank:
   (5/8) ____ (5/9)
   - Ⓐ >
   - Ⓑ <
   - Ⓒ =
   - Ⓓ none of these

3. The land area of Florida is 65,748 square miles. Write this area in expanded notation.
   - Ⓐ 6,000 + 500 + 70 + 48
   - Ⓑ 60,000 + 5,700 + 48
   - Ⓒ 60,000 + 5,000 + 700 + 40 + 8
   - Ⓓ 65,000 + 748

4. What is the value of $(3^2 + 6) \div 5$?
   - Ⓐ 5
   - Ⓑ 6
   - Ⓒ 3
   - Ⓓ 12

5. Estimate the product.
   831 × 96
   - Ⓐ 70,000
   - Ⓒ 79,776
   - Ⓑ 72,000
   - Ⓓ 80,000

6. What is the rule for the pattern?
   0, 3, 8, 15, 24, 35, ...
   for the nth term:
   - Ⓐ $n + 2$
   - Ⓑ $2n - 1$
   - Ⓒ $n^2 + 1$
   - Ⓓ $n^2 - 1$

7. What are the odds for winning 1 trip to a theme park in which 5,000 chances are given?
   - Ⓐ 1 in 4,999
   - Ⓑ 1 in 5,000
   - Ⓒ 4,999 in 1
   - Ⓓ 5,000 in 1

8. On a sunny day, the flagpole at Linda's school casts a 25 m long shadow. If Linda is 1.5 m tall and casts a shadow of 5 m at the same time of day, what is the height of the flagpole?
   - Ⓐ 7.5 m
   - Ⓒ 75 m
   - Ⓑ 37.5 m
   - Ⓓ 83 m

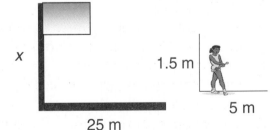

x 1.5 m

                           5 m

25 m

Holt Middle School Math   Course 1

End-of-Grade Practice Test 1

9. The length of a movie is 183 minutes. About how many hours long is the movie?

(A) 2 hours

(B) $2\frac{1}{2}$ hours

(C) 3 hours

(D) $3\frac{1}{2}$ hours

10. Evaluate:

$48 - 4 \times (2 + 6)$

(A) 12

(B) -32

(C) 32

(D) 16

11. A bottle-nosed dolphin can grow to be over 3 m long and weigh 200 kg. What is another way to write the weight of a bottle-nosed dolphin?

(A) 200 lb

(B) 2000 g

(C) 20,000 g

(D) 200,000 g

12. Which angle would be supplementary to a right angle?

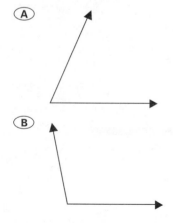

13. Which number is equal to 7.5%?

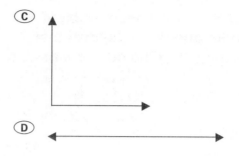

(A) $\frac{3}{4}$

(B) 0.075

(C) 0.75

(D) 7.5

14. State the best rule for the pattern.

2	4	8	16	32	64	128

(A) The numbers are increasing by 2.

(B) The numbers are increasing by 4.

(C) The numbers are doubling.

(D) The numbers are tripling.

15. What two numbers come next in the pattern?

1, 3, 6, 10, 15, 21, …

(A) 25, 33

(B) 26, 34

(C) 27, 35

(D) 28, 36

Holt Middle School Math   Course 1

End-of-Grade Practice Test 1

16. Use the stem-and-leaf plot of last month's daily temperatures in Tampa to determine how many days the temperature was less than 72°.

Stems	Leaves
6	2 5 6 6 8
7	0 2 3 6
8	2 5 7
9	0 5
10	0

  Ⓐ 4   Ⓒ 0

  Ⓑ 2   Ⓓ 6

17. Sam's bicycle tire has a radius of 35 cm. What is the circumference of the tire? Use 3.14 for π.

  Ⓐ 109.9 cm^2 Ⓒ 219.8 cm^2

  Ⓑ 109.9 cm   Ⓓ 219.8 cm

18. A local country music station polled 2,000 listeners to find out what type of music people preferred. The station reported that less than one third of the people enjoyed light rock. Why is this survey misleading?

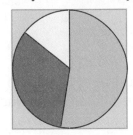

▨	Country
▩	Light Rock
▢	Instrumental

  Ⓐ The station only polled current listeners who like country music.

  Ⓑ They did not poll enough people.

  Ⓒ Most people like country music.

  Ⓓ The graph is not misleading.

19. Which choice represents the prime factorization of 90?

  Ⓐ 2 • 5 • 7

  Ⓑ 3 • 3 • 7

  Ⓒ 2 • 3 • 3 • 5

  Ⓓ 3 • 5 • 5

20. Which rule generates the table below?

  Ⓐ q = 3p + 2

  Ⓑ q = 2p + 2

  Ⓒ q = 3p

  Ⓓ q + 2 = 5p

p	q
0	2
1	5
2	8
3	11

21. In which quadrant does the point (3, –2) lie?

  Ⓐ I

  Ⓑ II

  Ⓒ III

  Ⓓ IV

22. Consider the following sets of numbers represent test scores for Mrs. Pinkerton's 3rd and 4th period classes:

3rd: 74, 78, 83, 86, 86, 88, 95
4th: 71, 79, 84, 88, 88, 90, 93, 100

Which of the following statements is true?

  Ⓐ 3rd period has a higher mode

  Ⓑ 4th period has a higher average

  Ⓒ 3rd period has a higher median

  Ⓓ none of these

103
Holt Middle School Math Course 1

End-of-Grade Practice Test 1

23. Mickey gets caught stealing base about 1 time out of every 25 attempts. If Mickey tries to steal base 350 times, how many times will he be called out?

Ⓐ 14

Ⓑ 24

Ⓒ 88

Ⓓ 336

24. Each student in Mrs. Hill's math class rolled a number cube 50 times and graphed the number of times each number came up. Which of the following most likely represents the graph the students made of the results of their whole class?

Ⓐ
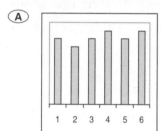

Ⓑ
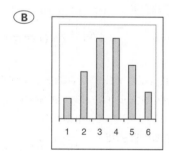

Ⓒ
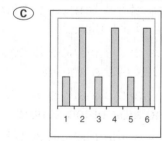

Ⓓ
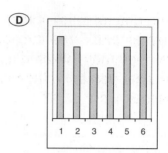

25. Nedda and Violetta are comparing their hat collections. Nedda has 4 more than three times as many hats Violetta. Which equation represents this if N = Nedda's hats and V = Violetta's hats?

Ⓐ $V = 3N + 4$

Ⓑ $V = 4N + 3$

Ⓒ $N = 4V + 3$

Ⓓ $N = 3V + 4$

26. A basketball team is sponsoring a car wash. The team can wash 5 cars in 75 minutes. How many cars can the team wash in 4 hours?

Ⓐ 12 cars

Ⓑ 14 cars

Ⓒ 16 cars

Ⓓ 18 cars

27. Shawn bought a blazer for 15% off. If the original price of the blazer was $35, how much did Shawn save?

Ⓐ $29.75

Ⓑ $20.00

Ⓒ $15.00

Ⓓ $ 5.25

Holt Middle School Math   Course 1

End-of-Grade Practice Test 1

28. The following graphs represent cars, vans, and trucks that came through a car wash during a day. Which graph best shows what percent o the total vehicles washed were cars?

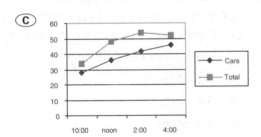

Ⓐ

Ⓑ

Ⓒ

Ⓓ

29. The dimensions of a given trapezoid are prime numbers. If the perimeter of the figure is 25 units, what are the possible dimensions?

Ⓐ 2, 5, 5, and 13
Ⓑ 3, 7, 7, and 8
Ⓒ 5, 5, 5, and 9
Ⓓ 3, 4, 9, and 9

For questions 30 and 31, use the picture below.

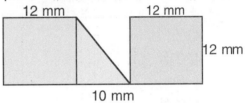

30. Find the area of the triangle.
Ⓐ 60 mm²   Ⓒ 144 mm²
Ⓑ 120 mm²   Ⓓ 1440 mm²

31. The triangle and each of the squares shown have the same base and height. What is the relationship between their areas?

Ⓐ The area of the square is 4 times the area of the triangle.

Ⓑ The area of the triangle is equal to the area of the square.

Ⓒ The area of the triangle is 1.5 times the area of the square.

Ⓓ The area of the triangle is 1/2 the area of the square.

Holt Middle School Math   Course 1

End-of-Grade Practice Test 1

32. An oak tree at the back of a house is $15\frac{3}{4}$ feet taller than the house. If the house is $62\frac{1}{2}$ feet at its peak, how tall is the oak tree?

Ⓐ $46\frac{3}{4}$ feet

Ⓑ $50\frac{1}{4}$ feet

Ⓒ $68\frac{3}{4}$ feet

Ⓓ $78\frac{1}{4}$ feet

33. Laura's baby sister weighed 4.2 kg at birth. How many grams did she weigh?

Ⓐ 42 g

Ⓑ 420 g

Ⓒ 4,200 g

Ⓓ 42,000 g

34. Given the ordered pairs below, what rule can be used to generate them?
(0, –5) , (2, –3), (3, –2), (5, 0)

Ⓐ $y = x - 5$

Ⓑ $y = x - 3$

Ⓒ $y = 2x - 5$

Ⓓ $x = 5x$

35. What is the reflection of the point (1, 2) about the line $x = 4$?

Ⓐ (2, –4)

Ⓑ (7, 2)

Ⓒ (2, 1)

Ⓓ (5, 2)

36. Maggie needs to find two congruent tile pieces for her art project. Which two pieces should she use?

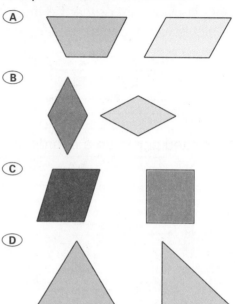

37. The rectangles in which pair are similar?

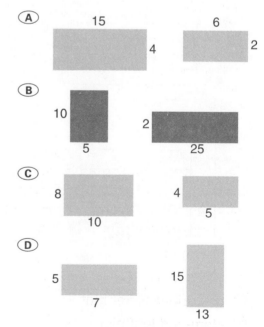

Holt Middle School Math   Course 1

End-of-Grade Practice Test 1

38. Which is the correct order of the following integers from GREATEST to LEAST?
4888, 4235, 4219, 4867, 3142

  Ⓐ 3142, 4219, 4235, 4867, 4888

  Ⓑ 4888, 4867, 4219, 4235, 3142

  Ⓒ 4219, 4867, 4888, 3142, 4235

  Ⓓ 4888, 4867, 4235, 4219, 3142

39. Mia plotted points on a coordinate plane. Which point is in Quadrant II?

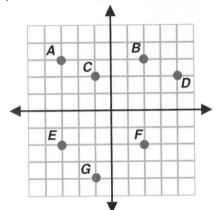

  Ⓐ C

  Ⓑ E

  Ⓒ F

  Ⓓ B

40. Mike and Matt collect baseball cards. They want to organize their cards in order to see how many of each card they have. Which type of graph would best display their data?

  Ⓐ bar graph

  Ⓑ line graph

  Ⓒ box-and-whisker plot

  Ⓓ stem-and-leaf plot

41. The results of the past five soccer seasons are graphed below. Which statement best describes the graph?

SOCCER STATS

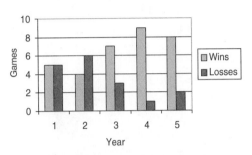

  Ⓐ The team has lost more games than it has won.

  Ⓑ The team is improving.

  Ⓒ The team's best season was the last season.

  Ⓓ Few people came to the games.

42. Which ordered pair is a solution to the equation $y = 3x + 2$?

  Ⓐ (2, 4)

  Ⓑ (1, 5)

  Ⓒ (26, 8)

  Ⓓ (1, 0)

43. The amount of money Maria has is about five times greater than the amount of money John has. Which equation best describes the function when y is equal to the amount of money Maria has?

  Ⓐ $5y = x$

  Ⓑ $y = x^5$

  Ⓒ $y = \dfrac{1}{5}x$

  Ⓓ $y = 5x$

Holt Middle School Math   **Course 1**

End-of-Grade Practice Test 1

44. Sandi used 20% of her savings to buy a bicycle. If the bike cost $125, how much money was originally in Sandi's savings account?

- Ⓐ $6.25
- Ⓑ $250
- Ⓒ $625
- Ⓓ $2,500

45. The dimensions of a given rectangle are both composite numbers. If the area of a rectangle is 40 square units, what are the dimensions of the rectangle?

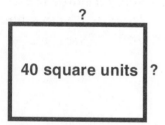

?

| 40 square units | ? |

- Ⓐ 1 and 40
- Ⓑ 2 and 20
- Ⓒ 4 and 10
- Ⓓ 5 and 8

46. When James entered the sixth grade, he was 62 inches tall, which is $5\frac{1}{4}$ inches taller than he was when he entered fifth grade. How tall was James when he entered the fifth grade?

- Ⓐ $56\frac{3}{4}$ inches
- Ⓑ $57\frac{1}{4}$ inches
- Ⓒ $57\frac{3}{4}$ inches
- Ⓓ $67\frac{1}{4}$ inches

47. Find the area of the shaded region.

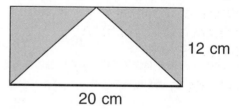

12 cm

20 cm

- Ⓐ 32 cm²
- Ⓑ 64 cm²
- Ⓒ 120 cm²
- Ⓓ 240 cm²

Holt Middle School Math   Course 1

End-of-Grade Practice Test 2

Directions: Read each problem carefully. Then, fill in the oval corresponding to the correct answer.

1. Jamie was in charge of bringing snacks to the soccer game. She brought 8 oranges, 3 apples, and 5 bananas. What fraction of the snacks were oranges?

   Ⓐ $\frac{1}{2}$ Ⓒ $\frac{3}{8}$

   Ⓑ $\frac{1}{3}$ Ⓓ $\frac{8}{11}$

2. It took Simon $2\frac{1}{2}$ hours to clean the pool. He spent $\frac{1}{3}$ of the time filling the pool with clean water. How much time did it take Simon to fill the pool?

   Ⓐ 30 minutes
   Ⓑ 40 minutes
   Ⓒ 50 minutes
   Ⓓ 60 minutes

3. Which of the following is in the correct order from LEAST to GREATEST?

   Ⓐ 3%, 0.27, 1/4

   Ⓑ 25%, 1/4, 1/5

   Ⓒ 1/5, 1/6, 1/7

   Ⓓ 5%, 1/3, 0.45

4. Barry has measured the length of a dollhouse windowsill as three inches. Express the length in feet.

   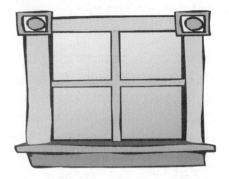

   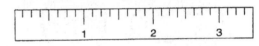

   Ⓐ $\frac{1}{2}$ ft   Ⓒ 0.15 ft

   Ⓑ $\frac{3}{4}$ ft   Ⓓ $\frac{1}{4}$ ft

5. Use the stem-and-leaf plot of last month's daily temperatures in Fort Lauderdale to determine how many days the temperature was over 85°.

Stems	Leaves
6	1 3 4
7	0 4 6
8	0 4 4 6 9
9	0 1 2 4 4 5
10	0 1 5

   Ⓐ 2 days   Ⓒ 9 days
   Ⓑ 11 days   Ⓓ 14 days

Holt Middle School Math   Course 1

End-of-Grade Practice Test 2

6. Evaluate the expression $8m - 4s$ for $m = 4$ and $s = 2$.

   Ⓐ 8   Ⓒ 28

   Ⓑ 24   Ⓓ 36

7. Sabrina is a dog groomer. She charges $15 to groom a dog plus $3 for every pound over 20 pounds that the dog weighs. Write an algebraic expression to describe Sabrina's pay rate when she grooms a dog that weighs over 20 pounds.

   Ⓐ $15w + 3$

   Ⓑ $15 + 3(20w)$

   Ⓒ $15 + 3w$

   Ⓓ $15 + 3(w - 20)$

8. Fran went to the store to buy school supplies. She bought a ruler for $1.35, a package of pencils for $2.07, and a box of crayons for $3.75. Which is the best estimate of how much she spent?

   Ⓐ $5.50   Ⓒ $6.25

   Ⓑ $6.75   Ⓓ $7.00

9. What are the coordinates of the reflection of point M about the line $x = 3$?

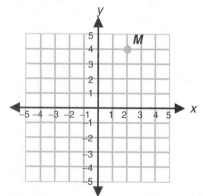

   Ⓐ (3, 3)   Ⓒ (2, 2)

   Ⓑ (3, 4)   Ⓓ (4, 4)

10. The hands on the clock form a 100° angle. Which of the following angles is supplementary to the angle on the clock?

   Ⓐ 260° angle

   Ⓑ 90° angle

   Ⓒ 80° angle

   Ⓓ 45° angle

11. For which problem would an estimate be a sufficient solution?

   Ⓐ If movie tickets sell for $8 a ticket, how much money do you need to buy 3 tickets?

   Ⓑ What is the maximum number of students allowed by law on a school bus?

   Ⓒ How much flour do you need to make a loaf of banana bread?

   Ⓓ How many guests are coming to the graduation party?

Holt Middle School Math   **Course 1**

End-of-Grade Practice Test 2

12. Megan has a fence around her backyard. Each fence post is 9 ft apart. What is the perimeter of Megan's backyard expressed in YARDS?

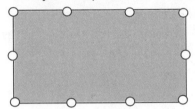

- Ⓐ 30 yds
- Ⓑ 60 yds
- Ⓒ 90 yds
- Ⓓ 100 yds

13. Fill in the blank:

9/8 ___ 8/9

- Ⓐ >
- Ⓑ <
- Ⓒ =
- Ⓓ none of these

14. Mr. Schmidt calculated the grades for his math class. What is the median of the grades?

85, 91, 92, 73, 74, 78, 86, 91, 94, 91, 87

- Ⓐ 85
- Ⓒ 87
- Ⓑ 86
- Ⓓ 91

15. What are the next two terms in the pattern?

5, 6, 9, 10, 13, 14, ...

- Ⓐ 15, 16
- Ⓑ 16, 17
- Ⓒ 16, 16
- Ⓓ 17, 18

16. Benjamin caught a grouper that weighed 1.4 kg. What is its weight in grams?

- Ⓐ 0.014 g
- Ⓑ 0.14 g
- Ⓒ 1,400 g
- Ⓓ 14,000 g

17. The chart shows the attendance at a college football team's first four games of the season.

Game 1	Game 2	Game 3	Game 4
48,785	32,152	45,321	39,421

Which computation method would you use to find the average attendance at the college football team's first four games of the season?

- Ⓐ paper and pencil
- Ⓑ mental math
- Ⓒ concrete materials
- Ⓓ calculator

18. Which of the following ratios is equal to the constant π? Hint: Think about the formulas you know related to circles.

- Ⓐ a circle's circumference to its diameter
- Ⓑ a circle's radius to its area
- Ⓒ a circle's area to its radius
- Ⓓ a circle's radius to its diameter

19. What is the approximate area of a circle with radius of length 3cm?

- Ⓐ 9.42 cm²
- Ⓒ 18.84 cm²
- Ⓑ 28.26 cm²
- Ⓓ 21.14 cm²

Holt Middle School Math   Course 1

End-of-Grade Practice Test 2

20. Identify the transformation shown below.

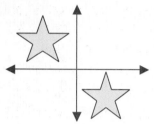

- Ⓐ reflection
- Ⓑ rotation
- Ⓒ translation
- Ⓓ dilation

21. At lunch Emma has three different vegetable options: carrots, peas, and green beans. She also has three drink options: soda, milk, and orange juice. In all, Emma has 9 possible combinations of a vegetable and a drink she can make. What is the probability that she'll pick green beans and milk?

- Ⓐ 1 in 18
- Ⓒ 1 in 9
- Ⓑ 1 in 3
- Ⓓ 1 in 6

22. It takes Ed $2\frac{1}{2}$ hr to mow his lawn,

$\frac{3}{4}$ hr to weed the flower bed,

and $\frac{1}{3}$ hr to water the flowers.

How much time does it take Ed to complete his yard work?

- Ⓐ $2\frac{5}{9}$ hr
- Ⓒ $2\frac{19}{12}$ hr
- Ⓑ $2\frac{7}{12}$ hr
- Ⓓ $3\frac{7}{12}$ hr

23. Select the Venn diagram that represents this situation. After interviewing 25 of her classmates, Jen found that 8 students owned dogs, 9 owned cats, 3 owned birds, and 10 did not own any of the three animals.

Ⓐ
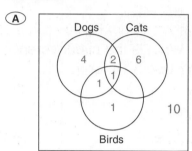

Ⓑ
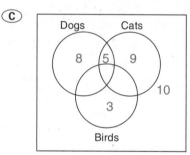

Ⓒ

Ⓓ
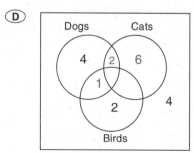

Holt Middle School Math   Course 1

End-of-Grade Practice Test 2

24. Compare:
7 ft ___ 6 yd
- (A) <
- (B) >
- (C) =
- (D) ≈

25. The basketball rim has a radius of 8 inches. What is the circumference of the rim? Use 3.14 for π.
- (A) 25.12 in
- (B) 25.12 in²
- (C) 50.24 in
- (D) 50.24 in²

26. What is the value of the following expression?
$$\frac{8 + 5}{6 + 10 \div 2 \times 4}$$
- (A) $\frac{1}{4}$
- (B) $\frac{13}{32}$
- (C) $\frac{1}{2}$
- (D) $\frac{13}{25}$

27. Which of the following correctly shows the prime factorization of the number 84?
- (A) $2^3 \cdot 3$
- (B) $2^2 \cdot 3 \cdot 7$
- (C) $2^3 \cdot 21$
- (D) $2^2 \cdot 3^2$

28. After a graduation party, the only food that was left was one-half of a 6-foot submarine sandwich. Brenda, Bob, Bill, and Beth decided to split the remainder between them. If Brenda took home 10% of the remaining sandwich, Bob took home $\frac{1}{4}$ of the remaining sandwich, and Bill took home $\frac{3}{5}$ of the remaining sandwich, how much of the remaining half-sandwich did Beth take home?
- (A) 5% of the sandwich
- (B) 10% of the sandwich
- (C) 15% of the sandwich
- (D) 20% of the sandwich

29. Estimate the area of the shaded region.

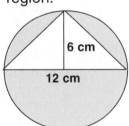

- (A) 149 cm²
- (B) 109 cm²
- (C) 77 cm²
- (D) 416 cm²

30. Which ordered pair is not a solution of the equation $y = 3x^2 + 1$?
- (A) (1, 4)
- (B) (-1, 4)
- (C) (0, -1)
- (D) (2, 13)

Holt Middle School Math   Course 1

End-of-Grade Practice Test 2

31. Micah needs to double his grand-mother's banana bread recipe. If he uses the recipe below, how much more flour will Micah need if he has only $3\frac{1}{2}$ cups of flour in his cabinet?

> ### *Banana Bread*
>
> 3 mashed bananas
>
> $2\frac{3}{4}$ cups of flour
>
> $\frac{3}{4}$ teaspoon vanilla
>
> $\frac{3}{4}$ teaspoon baking soda
>
> $\frac{3}{4}$ teaspoon salt

(A) $8\frac{1}{4}$ cups

(B) $4\frac{3}{4}$ cups

(C) $4\frac{1}{4}$ cups

(D) 2 cups

32. Beth and Barb babysit part-time. They charge $5.50 per hour plus $2 per child. Which expression can be use to show how much the girls are paid for watching *c* children for *h* hours?

(A) $5.5c + 2h$

(B) $2c + 5.5h$

(C) $10.1ch$

(D) $5.5h - 2c$

33. Sasha needs to find the height of the largest tree in her yard for a science project. She knows that the length of the shadow of her 3-ft tall mailbox is 4 ft long. If she measures the length of the shadow of the tree to be 8 yd at the same time, what is the height of Sasha's tree?

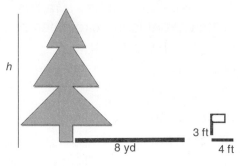

(A) 21 ft

(B) 18 ft

(C) 12 ft

(D) 6 ft

34. What is the sixth arrangement in this pattern?

(A)

(B)

(C)

(D)

Holt Middle School Math   Course 1

Name _____ Date _____ Class _____

End-of-Grade Practice Test 2

35. Point O is a reflection of point M across which line?

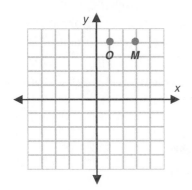

- Ⓐ $x = 2$
- Ⓑ $y = 2$
- Ⓒ $x = 3$
- Ⓓ $y = 2x + 1$

36. What relationship describes $\angle 4$ and $\angle 8$?

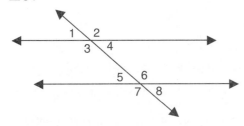

- Ⓐ Alternate Interior Angles
- Ⓑ Similar Angles
- Ⓒ Complimentary Angles
- Ⓓ Supplementary Angles

37. Which of the following figures is similar to the given figure?

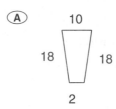

Ⓐ

10
18   18
2

Ⓑ

8
24   24
64

Ⓒ

64
12   12
16

Ⓓ

16
3   3
2

Holt Middle School Math   Course 1

End-of-Grade Practice Test 2

38. The number of people who rented a golf cart each day over the past three weeks is recorded in the stem-and-leaf plot. On how many days did at least 15 people rent a golf cart?

Stems	Leaves
0	5 6 7 8
1	5 5 5 6 7 8 9
2	1 1 2 3 3 6 8 8 9
3	2

- Ⓐ 21 days
- Ⓑ 17 days
- Ⓒ 14 days
- Ⓓ 8 days

39. In which quadrant would the point (–5, –3) appear on a coordinate plane?

- Ⓐ I
- Ⓒ III
- Ⓑ II
- Ⓓ IV

40. Examine the table below and determine the linear equation that describes the relationship between G and H.

G	H
–1	–6
0	–2
1	2
2	6
3	10

- Ⓐ G = –4H – 2
- Ⓑ H = 4G – 2
- Ⓒ G = 4H – 2
- Ⓓ H = 2G – 4

41. Charlie works at a grocery store. He earns $7.25 per hour. He works 25 hours each week and is paid weekly. If 18.75% of his earnings are withheld for taxes and he puts 10% of his remaining pay into a savings account, how long will it take Charlie to save $100?

- Ⓐ 5 weeks
- Ⓑ 6 weeks
- Ⓒ 7 weeks
- Ⓓ 8 weeks

42. Which of the following correctly shows the prime factorization of the number 600?

- Ⓐ $2^3 \times 3 \times 5^2$
- Ⓑ $2^3 \times 3^2 \times 5^2$
- Ⓒ $2^2 \times 3^3 \times 5^2$
- Ⓓ $2^2 \times 3^2 \times 5^3$

43. A certain recipe calls for 3 cups of flour, 6 eggs, $\frac{3}{4}$ cup of sugar, and 1 teaspoon of salt. If Sue has only 1 egg, how much sugar does she need if she adjusts the recipe accordingly?

- Ⓐ $\frac{1}{4}$ cup
- Ⓑ $\frac{1}{6}$ cup
- Ⓒ $\frac{1}{8}$ cup
- Ⓓ $\frac{1}{10}$ cup

Holt Middle School Math   Course 1

End-of-Grade Practice Test 1 Answer Key and Correlations

Item	Answer	State of North Carolina Mathematics Standards, Grade 6
1	B	2.01 Estimate and measure length, perimeter, area, angles, weight, and mass of two-and three-dimensional figures, using appropriate tools.
2	A	1.03 Compare and order rational numbers.
3	C	1.05 Develop fluency in the use of factors, multiples, exponential notation, and prime factorization.
4	C	1.05 Develop fluency in the use of factors, multiples, exponential notation, and prime factorization.
5	D	1.04c Develop fluency in addition, subtraction, multiplication and division of non-negative rational numbers. Estimate the results of computations
6	D	1.03 Compare and order rational numbers.
7	B	4.02 Use a sample space to determine the probability of an event.
8	A	Gr. 7, 1.01 Develop and use ratios, proportions, and percents to solve problems. Gr. 7, 2.01 Draw objects to scale and use scale drawings to solve problems.
9	C	N/A
10	D	5.01e Simplify algebraic expressions and verify the results using the basic properties of rational numbers. Order of operations.
11	D	2.01 Estimate and measure length, perimeter, area, angles, weight, and mass of two-and three-dimensional figures, using appropriate tools.
12	C	2.01 Estimate and measure length, perimeter, area, angles, weight, and mass of two-and three-dimensional figures, using appropriate tools.
13	B	1.02a Develop meaning for percents. Connect the model, number word, and number using a variety of representations. 1.03 Compare and order rational numbers.
14	C	1.03 Compare and order rational numbers.
15	D	1.03 Compare and order rational numbers.
16	C	Gr. 7, 4.01 Collect, organize, analyze, and display data (including box plots and histograms) to solve problems.
17	D	2.02 Solve problems involving perimeter/circumference and area of plane figures.
18	A	4.06 Design and conduct experiments or surveys to solve problems; report and analyze results.
19	C	1.05 Develop fluency in the use of factors, multiples, exponential notation, and prime factorization.
20	A	5.04 Use graphs, tables, and symbols to model and solve problems involving rates of change and ratios.
21	D	3.04 Solve problems involving geometric figures in the coordinate plane.
22	B	Gr. 7, 4.02 Calculate, use, and interpret the mean, median, mode, range, frequency distribution, and inter-quartile range for a set of data.
23	A	4.02 Use a sample space to determine the probability of an event.
24	A	4.06 Design and conduct experiments or surveys to solve problems; report and analyze results.
25	D	5.02 Use and evaluate algebraic expressions.
26	C	Gr. 7, 1.01 Develop and use ratios, proportions, and percents to solve problems.
27	D	Gr. 7, 1.01 Develop and use ratios, proportions, and percents to solve problems.
28	B	Gr. 7, 4.01 Collect, organize, analyze, and display data (including box plots and histograms) to solve problems.
29	A	2.02 Solve problems involving perimeter/circumference and area of plane figures.
30	A	2.02 Solve problems involving perimeter/circumference and area of plane figures.

31	D	2.02 Solve problems involving perimeter/circumference and area of plane figures.
32	D	1.04 Develop fluency in addition, subtraction, multiplication, and division of non-negative rational numbers. 1.05 Develop fluency in the use of factors, multiples, exponential notation, and prime factorization.
33	C	2.01 Estimate and measure length, perimeter, area, angles, weight, and mass of two-and three-dimensional figures, using appropriate tools.
34	A	Gr. 7, 5.01 Identify, analyze, and create linear relations, sequences, and functions using symbols, graphs, tables, diagrams, and written descriptions.
35	B	3.03 Transform figures in the coordinate plane and describe the transformation.
36	B	Gr. 7, 3.03 Use scaling and proportional reasoning to solve problems related to similar and congruent polygons.
37	C	Gr. 7, 3.03 Use scaling and proportional reasoning to solve problems related to similar and congruent polygons.
38	D	1.03 Compare and order rational numbers.
39	A	3.04 Solve problems involving geometric figures in the coordinate plane.
40	A	Gr. 7, 4.01 Collect, organize, analyze, and display data (including box plots and histograms) to solve problems.
41	B	Gr. 7, 4.01 Collect, organize, analyze, and display data (including box plots and histograms) to solve problems.
42	B	Gr. 7, 5.03 Use and evaluate algebraic expressions, linear equations or inequalities to solve problems.
43	D	5.02 Use and evaluate algebraic expressions.
44	C	Gr. 7, 1.01 Develop and use ratios, proportions, and percents to solve problems.
45	C	2.02 Solve problems involving perimeter/circumference and area of plane figures.
46	A	1.04 Develop fluency in addition, subtraction, multiplication, and division of non-negative rational numbers. 1.05 Develop fluency in the use of factors, multiples, exponential notation, and prime factorization. 2.01 Estimate and measure length, perimeter, area, angles, weight, and mass of two-and three-dimensional figures, using appropriate tools.
47	C	2.02 Solve problems involving perimeter/circumference and area of plane figures.

End-of-Grade Practice Test 2 Answer Key and Correlations

Item	Answer	State of North Carolina Mathematics Standards, Grade 6
1	A	1.05 Develop fluency in the use of factors, multiples, exponential notation, and prime factorization.
2	C	1.05 Develop fluency in the use of factors, multiples, exponential notation, and prime factorization.
3	D	1.03 Compare and order rational numbers. 1.05 Develop fluency in the use of factors, multiples, exponential notation, and prime factorization.
4	D	2.01 Estimate and measure length, perimeter, area, angles, weight, and mass of two-and three-dimensional figures, using appropriate tools.
5	B	Gr. 7, 4.01 Collect, organize, analyze, and display data (including box plots and histograms) to solve problems.
6	B	5.02 Use and evaluate algebraic expressions.
7	D	5.02 Use and evaluate algebraic expressions.
8	C	1.04c Develop fluency in addition, subtraction, multiplication, and division of non-negative rational numbers. Estimate the results of computations.
9	D	3.04 Solve problems involving geometric figures in the coordinate plane.
10	C	2.01 Estimate and measure length, perimeter, area, angles, weight, and mass of two-and three-dimensional figures, using appropriate tools.
11	D	1.04c Develop fluency in addition, subtraction, multiplication, and division of non-negative rational numbers. Estimate the results of computations.
12	A	2.02 Solve problems involving perimeter/circumference and area of plane figures.
13	A	1.03 Compare and order rational numbers. 1.05 Develop fluency in the use of factors, multiples, exponential notation, and prime factorization.
14	C	Gr.7, 4.02 Calculate, use, and interpret the mean, median, mode, range, frequency distribution, and inter-quartile range for a set of data.
15	D	1.03 Compare and order rational numbers.
16	C	2.01 Estimate and measure length, perimeter, area, angles, weight, and mass of two-and three-dimensional figures, using appropriate tools.
17	D	1.07 Develop flexibility in solving problems by selecting strategies and using mental computation, estimation, calculators or computers, and paper and pencil.
18	A	3.02 Identify the radius, diameter, chord, center, and circumference of a circle; determine the relationships among them.
19	B	2.02 Solve problems involving perimeter/circumference and area of plane figures.
20	C	3.03 Transform figures in the coordinate plane and describe the transformation.
21	C	4.01 Develop fluency with counting strategies to determine the sample space for an event. Include lists, tree diagrams, frequency distribution tables, permutations, combinations, and the Fundamental Counting Principle.
22	D	1.04 Develop Fluency in addition, subtraction, multiplication, and division of non-negative rational numbers. 1.05 Develop fluency in the use of factors, multiples, exponential notation, and prime factorization.
23	A	1.07 Develop flexibility in solving problems by selecting strategies and using mental computation, estimation, calculators or computers, and paper and pencil. 4.06 Design and conduct experiments or surveys to solve problems; report and analyze results.
24	A	1.03 Compare and order rational numbers.
25	A	2.02 Solve problems involving perimeter/circumference and area of plane figures.

26	C	5.01e Simplify algebraic expressions and verify the results using the basic properties of rational numbers. Order of operations. 5.02 Use and evaluate algebraic expressions.
27	B	1.05 Develop fluency in the use of factors, multiples, exponential notation, and prime factorization.
28	A	1.02b Develop meaning for percents. Make estimates in appropriate situations. 1.05 Develop fluency in the use of factors, multiples, exponential notation, and prime factorization.
29	C	2.02 Solve problems involving perimeter/circumference and area of plane figures. 3.02 Identify the radius, diameter, chord, center, and circumference of a circle; determine the relationships among them.
30	C	Gr. 7, 5.01 Identify, analyze, and create linear relations, sequences, and functions using symbols, graphs, tables, diagrams, and written descriptions.
31	D	1.04 Develop Fluency in addition, subtraction, multiplication, and division of non-negative rational numbers. 1.05 Develop fluency in the use of factors, multiples, exponential notation, and prime factorization.
32	B	5.02 Use and evaluate algebraic expressions.
33	B	Gr. 7, 1.01 Develop and use ratios, proportions, and percents to solve problems.
34	C	Gr. 7, 5.01 Identify, analyze, and create linear relations, sequences, and functions using symbols, graphs, tables, diagrams, and written descriptions.
35	A	3.03 Transform figures in the coordinate plane and describe the transformation.
36	A	3.01 Identify and describe the intersection of figures in a plane.
37	D	Gr. 7, 3.02 Identify, define, and describe similar and congruent polygons with respect to angle measures, length of sides, and proportionality of sides. Gr. 7, 3.03 Use scaling and proportional reasoning to solve problems related to similar and congruent polygons.
38	B	Gr. 7, 4.01 Collect, organize, analyze, and display data (including box plots and histograms) to solve problems.
39	C	3.04 Solve problems involving geometric figures in the coordinate plane.
40	B	Gr. 7, 5.01 Identify, analyze, and create linear relations, sequences, and functions using symbols, graphs, tables, diagrams, and written descriptions.
41	C	1.02 Develop meaning for percents. Gr. 7, 1.01 Develop and use ratios, proportions, and percents to solve problems.
42	A	1.05 Develop fluency in the use of factors, multiples, exponential notation, and prime factorization.
43	C	1.05 Develop fluency in the use of factors, multiples, exponential notation, and prime factorization.